Polymer Materials and Their Applications

高分子材料与应用

王选伦　吕　军　编著

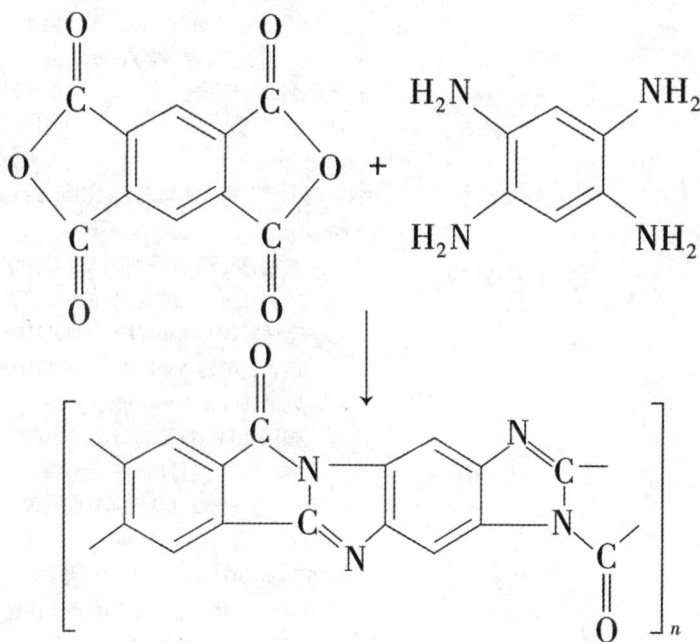

重庆大学出版社

内 容 提 要

  本书共 8 章,内容主要包括高分子材料的发展历史和分类,热塑性常用塑料、热固性塑料、工程塑料的制备方法,合成原理、品种、性能及用途,常见天然纤维和化学纤维的分类、制备原理、性能特点及用途,橡胶的常用品种、加工方法、配方体系、性能及用途,涂料及黏合剂的分类、性能及用途,一些功能高分子的性能特点、合成路线及用途等。

图书在版编目(CIP)数据

高分子材料与应用=Polymer Materials and Their
Applications:英文/王选伦,吕军编著.—重庆:重庆大
学出版社,2015.6(2017.1 重印)
  ISBN 978-7-5624-9153-8

  Ⅰ.①高…　Ⅱ.①王…②吕…　Ⅲ.①高分子材料—英文
Ⅳ.①TB324

  中国版本图书馆 CIP 数据核字(2015)第 128566 号

Polymer Materials and Their Applications
高分子材料与应用
GAOFENZI CAILIAO YU YINGYONG

王选伦　吕　军　编著
策划编辑:周　立
责任编辑:李定群　涂　昀　　版式设计:周　立
责任校对:贾　梅　　　　　　责任印制:赵　晟

*

重庆大学出版社出版发行
出版人:易树平
社址:重庆市沙坪坝区大学城西路 21 号
邮编:401331
电话:(023) 88617190　88617185(中小学)
传真:(023) 88617186　88617166
网址:http://www.cqup.com.cn
邮箱:fxk@ cqup.com.cn(营销中心)
全国新华书店经销
重庆学林建达印务有限公司印刷

*

开本:787mm×1092mm　1/16　印张:15.75　字数:373 千
2015 年 6 月第 1 版　　2017 年 1 月第 2 次印刷
ISBN 978-7-5624-9153-8　定价:39.80 元

编写说明

"高分子材料与应用"在国内很多高校是高分子材料与工程专业的一门专业必修课,有些同时是双语课。该课程旨在帮助高分子材料与工程专业的学生了解并掌握常见高分子材料的种类以及高分子材料的基本特性,使其具备高分子材料制备、结构与性能、成型加工、制品种类及应用领域等方面的基本知识。

这门课程理论不深,但教学内容广泛,实用性很强,很适合开设为双语课。很多高校的高分子材料专业开设了"高分子材料与应用"课程,但是有的不是双语课,就使用中文教材,是双语课的,也一般用自编讲义,没有合适的教材。在近5年的教学过程中,我们发现只使用中文教材,难以取得较好的教学效果。学生往往只看中文教材,而不去听教师的英文讲解,达不到提高学生阅读专业文献的能力和掌握专业词汇的目标。如使用英文原版教材,则存在与我国教学大纲不完全相符、内容过于全面且难度无法调节等问题。在实际讲课过程中,教师为了照顾到各个层次的学生和教学大纲的要求,往往对教材内容进行选择性讲解,实际教学内容覆盖率低,且学生学习起来知识点散,不易系统组织,同时也造成了教材的浪费。因此,编写一本符合国内需求的,内容比较丰富而实用的双语课教材是很有意义而且必要的。

本书的内容主要包括介绍高分子材料的发展历史和分类,重点讲解了通用塑料、工程塑料、热固性塑料的合成原理、常见品种、加工工艺、改性途径、主要性能及应用,介绍了常见天然纤维和化学纤维的分类、制备原理、性能特点及应用,橡胶的常见品种、加工方法、配方体系、性能及应用,涂料及黏合剂的分类、合成工艺、主要性能及应用,一些重要功能高分子材料的性能特点、合成路线及应用等。

本书每章节后有课后练习,练习答案在本书的课件中,课件可在重庆大学出版社的官网上下载。

在本书的编写过程中,得到了重庆理工大学材料学院杜长华院长的鼓励和支持。全书共分 8 章,第 1—6 章由王选伦编写,第 7,8 章由吕军编写。本书内容系统全面,围绕当前高分子材料的种类、性能、加工与应用以及技术发展均给出了详细的论述和介绍,不仅可作为高分子材料与工程相关专业的教材,也是材料工程技术人员的一本很好的参考书。

本书由重庆理工大学市级重点学科材料科学与工程,中国国家自然科学基金委员会(51373139)及聚合物分子工程国家重点实验室(复旦大学)开放研究课题基金(K2014-02)资助出版,在此表示感谢。

由于编者水平及时间所限,书中不足或谬误之处在所难免,恳请读者批评指正。

编 者

2015 年 1 月

CONT-ENTS

1

2

CHAPTER 1

INTRODUCTION

1.1 What are polymers?

Plastics, rubbers, fibers are all polymers. What is a polymer? The simplest definition of a polymer is something made of many units. Think of a polymer as a chain (Figure 1.1). The term *polymer* is derived from the Greek words *poly* and *meros*, meaning many parts[1]. Each link of the chain is the "mer" or basic unit that is usually made of carbon, hydrogen, oxygen, and/or silicon. To make the chain, many links or "mers" are hooked or polymerized together. Polymerization can be demonstrated by linking strips of construction paper together to make paper garlands or hooking together hundreds of paper clips to form chains.

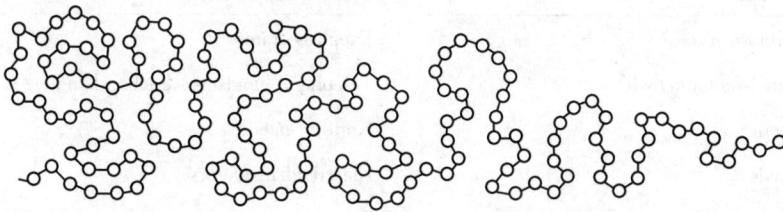

Figure 1.1　Model of a polymer molecule

Polymers have been with us since the beginning of time. Natural polymers include such things as tar and shellac, tortoise shell and horns, as well as tree saps that produce amber and latex. These polymers were processed with heat and pressure into useful articles like hair ornaments and jewelry. Natural polymers began to be chemically modified during the 1800s to produce many materials. The most famous of these were vulcanized rubber, gun cotton, and celluloid. The first semi-synthetic polymer produced was Bakelite in 1909 and was soon followed by the first synthetic fiber, rayon, which was developed in 1911.

Even with these developments, it was not until World War II that significant changes took place in the polymer industry. Prior to World War II, natural substances were generally available; therefore, synthetics that were being developed were not a necessity. Once the world went to war, our natural sources of latex, wool, silk, and other materials were cut off, making the use of synthetics critical. During this time period, we saw the use of nylon, acrylic, neoprene, SBR, polyethylene, and many more polymers take the place of natural materials that were no longer available. Since then, the polymer industry has continued to grow and has evolved into one of the fastest growing industries in the U.S. and in the world.

Table 1.1 gives a list of some polymers, their year of introduction, and some of their applications. It is obvious that the pace of development of plastics, which was painfully slow up to the 1920s, picked up considerable momentum in the 1930s and the 1940s. The first generation of man-made polymers was the result of empirical activities; the main focus was on chemical composition with virtually no attention paid to structure. However, during the first half of the 20th century, extensive organic and physical developments led to the first understanding of the structural concept of polymers—long chains or a network of covalently bonded molecules. In this regard the classic work of the German chemist Hermann Staudinger on polyoxymethylene and rubber and of the American chemist W. T. Carothers on nylon stand out clearly. Staudinger first proposed the theory that polymers were composed of giant molecules, and he coined the word *macromolecule* to describe them. Carothers discovered nylon, and his fundamental research (through which nylon was actually discovered) contributed considerably to the elucidation of the nature of polymers. His classification of polymers as *condensation* or *addition* polymers persists today.

Table 1.1 Brief history of polymeric materials

Date	Material	Typical Use
1868	Cellulose nitrate	Eyeglass frames
1909	Phenol-formaldehyde	Telephone handsets, knobs, handles
1919	Casein	Knitting needles
1926	Alkyds	Electrical insulators
1927	Cellulose acetate	Toothbrushes, packaging
1927	Poly(vinyl chloride)	Raincoats, flooring
1929	Urea-formaldehyde	Lighting fixtures, electrical switches
1935	Ethyl cellulose	Flashlight cases
1936	Polyacrylonitrile	Brush backs, displays
1936	Poly(vinyl acetate)	Flashbulb lining, adhesives
1938	Cellulose acetate butyrate	Irrigation pipe
1938	Polystyrene	Kitchenwares, toys
1938	Nylon (polyamide)	Gears, fibers, films

continued

Date	Material	Typical Use
1938	Poly(vinyl acetal)	Safety glass interlayer
1939	Poly(vinylidene chloride)	Auto seat covers, films, paper, coatings
1939	Melamine-formaldehyde	Tableware
1942	Polyester (cross-linkable)	Boat hulls
1942	Polyethylene (low density)	Squeezable bottles
1943	Fluoropolymers	Industrial gaskets, slip coatings
1943	Silicone	Rubber goods
1945	Cellulose propionate	Automatic pens and pencils
1947	Epoxies	Tools and jigs
1948	Acrylonitrile-butadiene-styrene copolymer	Luggage, radio and television cabinets
1949	Allylic	Electrical connectors
1954	Polyurethane	Foam cushions
1956	Acetal resin	Automotive parts
1957	Polypropylene	Safety helmets, carpet fiber
1957	Polycarbonate	Appliance parts
1959	Chlorinated polyether	Valves and fittings
1962	Phenoxy resin	Adhesives, coatings
1962	Polyallomer	Typewriter cases
1964	Ionomer resins	Skin packages, moldings
1964	Polyphenylene oxide	Battery cases, high temperature moldings
1964	Polyimide	Bearings, high temperature films and wire coatings
1964	Ethylene-vinyl acetate	Heavy gauge flexible sheeting
1965	Polybutene	Films
1965	Polysulfone	Electrical/electronic parts
1970	Thermoplastic polyester	Electrical/electronic parts
1971	Hydroxy acrylates	Contact lenses
1973	Polybutylene	Piping
1974	Aromatic polyamides	High-strength tire cord
1975	Nitrile barrier resins	Containers

The initial compound that is used to form polymers is the "mer" or monomer. Monomers are chemically joined together in one of two ways: addition polymerization or condensation polymerization.

Addition polymerization is comprised of three basic steps: initiation, propagation, and

3

termination. For example, during the initiation phase of the polymerization of polyethylene, the double bonds in the ethylene "mers" break and begin to bond together. A catalyst or promoter may be necessary to begin or speed up the reaction. The second phase, propagation, involves the continued addition of monomers together into chains.

The final step is termination. During termination all monomers may be used, causing the reaction to cease. A polymerization reaction can cease by quenching the reaction. Similar to quenching someone's thirst, water can be used to quickly cool a reaction. Polymers formed by addition polymerization include acrylic, polyethylene, and polystyrene, to name a few.

Very simply, addition polymerization describes the process of "mers" joining by each one adding on to the end of the last "mer". A simple visual of the process is paper clips joined together to form a long chain. Polymers formed by addition polymerization are often thermoplastic in nature. Thermoplastics are like hot melt glue sticks that can be heated and made soft and then become hard when cooled. Thermoplastic polymers are easily processed and reprocessed or recycled. The majority of polymers used today are thermoplastics.

The other group of polymers is formed by condensation polymerization. During the chemical reaction of condensation polymerization, a small molecule is eliminated as the monomers join together. Common polymers in this group include nylons, some polyesters, urea formaldehyde, and urethanes. These polymers can be thermoplastic in nature or thermosetting. Once a thermoset polymer is formed, it cannot be melted and reformed. All plastics flow at some time during their processing and are solid in the finished state, but once a thermoset is processed, it is dramatically different and cannot be reformed.

The means of polymerization will affect the heat reaction of the formed polymer; likewise, the arrangement of the "mers" within the molecule will affect the physical characteristics of the formed polymer. "Mers" joined together in long chains have a linear configuration very similar to a paper clip chain, even though in actuality tetrahedral bonds give the molecule a zigzag arrangement. During polymerization, if the "mers" not only form straight chains but also form long side chains off the main backbone, the resulting configuration is described as branched, like a tree branch or grape stem. A third configuration is achieved by the long chains being chemically linked together. An example would be natural rubber (isoprene) being reacted with sulfur. The sulfur bonds the chains to form a giant meshwork molecular structure that is known as vulcanized rubber. This is a cross-linked configuration.

Polyethylene has the simplest "mer" structure. Even though the backbone of other polymers will be similarly formed by a broken bond between two carbons, the remaining carbons in the "mer" will form a functional group whose orientation about the backbone will affect the physical nature of the resulting polymer. For example, propylene is the "mer" that will form polypropylene:

$$\begin{array}{cc} H & H \\ | & | \\ C & = C \\ | & | \\ H & CH_3 \end{array}$$

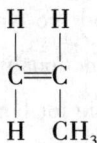

Polymerization will be initiated by the double bond breaking and the "-mers" joining together. Therefore, the methyl group on the propylene "-mer" has the potential to be located at various points along the backbone. If the methyl group (CH_3) is oriented repeatedly on one side of the chain on alternating carbons, it is called isotactic. Ninety to nine-five percent of all polypropylene polymers have this configuration.

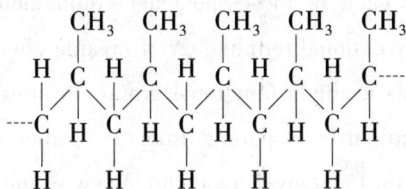

$$\begin{array}{ccccc} CH_3 & CH_3 & CH_3 & CH_3 & CH_3 \\ H \hspace{0.3em} C \hspace{0.3em} H & C \hspace{0.3em} H & C \hspace{0.3em} H & C \hspace{0.3em} H & C^{----} \\ | \diagup | \diagdown | & \diagup | \diagdown | & \diagup | \diagdown | & \diagup | \diagdown | & \diagup | \\ ^{----}C \hspace{0.3em} H & C \hspace{0.3em} H & C \hspace{0.3em} H & C \hspace{0.3em} H & C \hspace{0.3em} H \\ | & | & | & | & | \\ H & H & H & H & H \end{array}$$

1.2   Classification of polymers

Polymers can be classified in many different ways. The most obvious classification is based on the origin of the polymer, i.e., natural vs. synthetic. Other classifications are based on the polymer structure, polymerization mechanism, preparative techniques, or thermal behavior[2].

1.2.1   Natural vs. synthetic

Polymers may either be naturally occurring or purely synthetic. All the conversion processes occurring in our body (e.g., generation of energy from our food intake) are due to the presence of enzymes. Life itself may cease if there is a deficiency of these enzymes. Enzymes, nucleic acids, and proteins are polymers of biological origin. Their structures, which are normally very complex, were not understood until very recently. Starch—a staple food in most cultures—cellulose, and natural rubber, on the other hand, are examples of polymers of plant origin and have relatively simpler structures than those of enzymes or proteins. There are a large number of synthetic (man-made) polymers consisting of various families: fibers, elastomers, plastics, adhesives, etc. Each family itself has subgroups.

1.2.2   Polymer structure

(1) Linear, branched or cross-linked, ladder vs. functionality

As we stated earlier, a polymer is formed when a very large number of structural units (repeating units, monomers) are made to link up by covalent bonds under appropriate conditions. Certainly even if the conditions are "right" not all simple (small) organic molecules possess the

5

ability to form polymers. In order to understand the type of molecules that can form a polymer, let us introduce the term *functionality*. The functionality of a molecule is simply its interlinking capacity, or the number of sites it has available for bonding with other molecules under the specific polymerization conditions. A molecule may be classified as monofunctional, bifunctional, or polyfunctional depending on whether it has one, two, or greater than two sites available for linking with other molecules. For example, the extra pair of electrons in the double bond in the styrene molecules endows it with the ability to enter into the formation of two bonds. Styrene is therefore bifunctional. The presence of two condensable groups in both hexamethylenediamine ($-NH_2$) and adipic acid ($-COOH$) makes each of these monomers bifunctional. However, functionality as defined here differs from the conventional terminology of organic chemistry where, for example, the double bond in styrene represents a single functional group. Besides, even though the interlinking capacity of a monomer is ordinarily apparent from its structure, functionality as used in polymerization reactions is specific for a given reaction. A few examples will illustrate this.

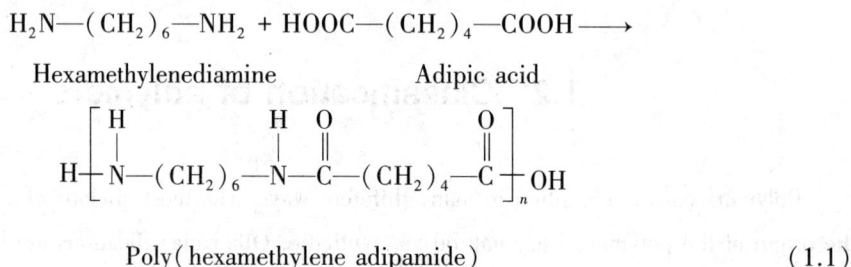

$$H_2N-(CH_2)_6-NH_2 + HOOC-(CH_2)_4-COOH \longrightarrow$$

Hexamethylenediamine   Adipic acid

$$H-\left[\underset{\overset{|}{H}}{N}-(CH_2)_6-\underset{\overset{|}{H}}{N}-\underset{\overset{||}{O}}{C}-(CH_2)_4-\underset{\overset{||}{O}}{C} \right]_n OH$$

Poly(hexamethylene adipamide)   (1.1)

A diamine like hexamethylenediamine has a functionality of 2 in amide-forming reactions such as that shown in Equation 1.1. However, in esterification reactions a diamine has a functionality of zero. Butadiene has the following structure:

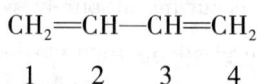

$$\underset{1\qquad 2\qquad 3\qquad 4}{CH_2=CH-CH=CH_2}$$

From our discussion about the polymerization of styrene, the presence of two double bonds on the structure of butadiene would be expected to prescribe a functionality of 4 for this molecule. Butadiene may indeed be tetra-functional, but it can also have a functionality of 2 depending on the reaction conditions (Equation 1.2).

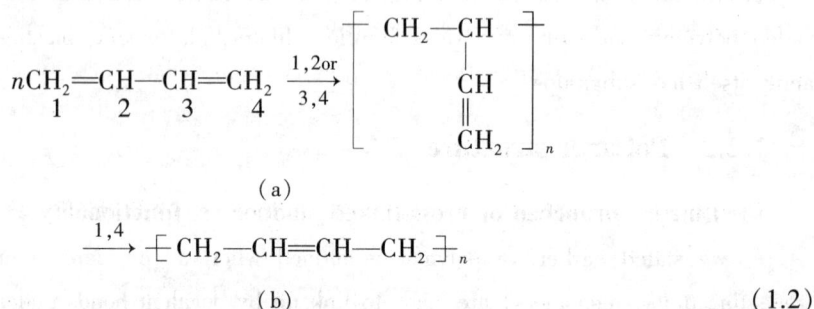

$$n\underset{1\quad 2\quad 3\quad 4}{CH_2=CH-CH=CH_2} \xrightarrow[3,4]{1,2\text{or}} \left[\begin{array}{c} CH_2-CH \\ | \\ CH \\ || \\ CH_2 \end{array} \right]_n$$

(a)

$$\xrightarrow{1,4} \left[CH_2-CH=CH-CH_2 \right]_n$$

(b)   (1.2)

Since there is no way of making a distinction between the 1,2 and 3,4 double bonds, the reaction of either double bond is the same. If either of these double bonds is involved in the polymerization reaction, the residual or unreacted double bond is on the structure attached to the main chain [i.e., part of the pendant group (a)]. In 1,4 polymerization, the residual double bond shifts to the 2,3 position along the main chain. In either case, the residual double bond is inert and is generally incapable of additional polymerization under the conditions leading to the formation of the polymer. In this case, butadiene has a functionality of 2. However, under appropriate reaction conditions such as high temperature or cross-linking reactions, the residual unsaturation either on the pendant group or on the backbone can undergo additional reaction. In that case, butadiene has a total functionality of 4 even though all the reactive sites may not be activated under the same conditions. Monomers containing functional groups that react under different conditions are said to possess latent functionality.

Now let us consider the reaction between two monofunctional monomers such as in an esterification reaction (Equation 1.3).

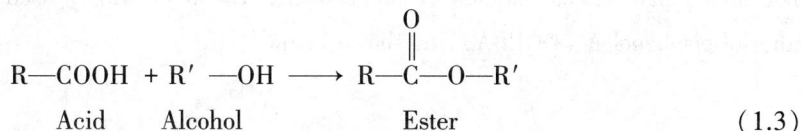

$$R-COOH + R'-OH \longrightarrow R-\overset{\overset{\displaystyle O}{\|}}{C}-O-R'$$

Acid　　　　Alcohol　　　　　　　Ester　　　　　　　　　　　(1.3)

You will observe that the reactive groups on the acid and alcohol are used up completely and that the product ester is incapable of further esterification reaction. But what happens when two bifunctional molecules react? Let us use esterification once again to illustrate the principle (Equation 1.4).

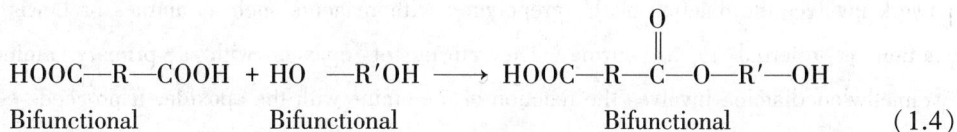

$$HOOC-R-COOH + HO-R'OH \longrightarrow HOOC-R-\overset{\overset{\displaystyle O}{\|}}{C}-O-R'-OH$$

Bifunctional　　　　　Bifunctional　　　　　　　　Bifunctional　　　　　　(1.4)

The ester resulting from this reaction is itself bifunctional, being terminated on either side by groups that are capable of further reaction. In other words, this process can be repeated almost indefinitely. The same argument holds for polyfunctional molecules. It is thus obvious that the generation of a polymer through the repetition of one or a few elementary units requires that the molecule(s) must be at least bifunctional.

The structural units resulting from the reaction of monomers may in principle be linked together in any conceivable pattern. Bifunctional structural units can enter into two and only two linkages with other structural units. This means that the sequence of linkages between bifunctional units is necessarily linear. The resulting polymer is said to be linear. However, the reaction between polyfunctional molecules results in structural units that may be linked so as to form nonlinear structures. In some cases the side growth of each polymer chain may be terminated before the chain has a chance to link up with another chain. The resulting polymer molecules are said to be branched. In other cases, growing polymer chains become chemically linked to each other, resulting

7

in a cross-linked system (Figure 1.2).

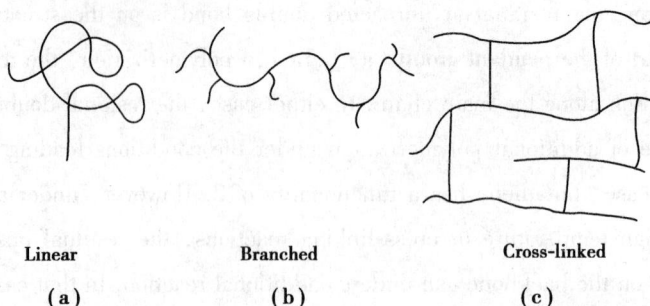

Linear　　　　　**Branched**　　　　　**Cross-linked**

(a)　　　　　　　**(b)**　　　　　　　**(c)**

Figure 1.2　Linear, branched, and cross-linked polymers

The formation of a cross-linked polymer is exemplified by the reaction of epoxy polymers, which have been used traditionally as adhesives and coatings and, more recently, as the most common matrix in aerospace composite materials. Epoxies exist at ordinary temperatures as low-molecular-weight viscous liquids or prepolymers. The most widely used prepolymer is diglycidyl ether of bisphenol A (DGEBA), as shown below:

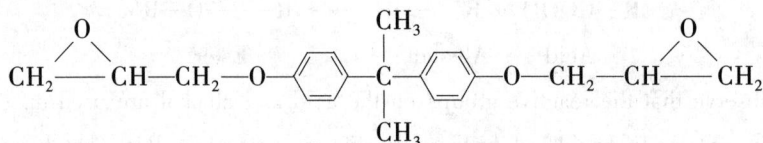

$$CH_2\overset{O}{-\!\!\triangle\!\!-}CH-CH_2-O-\!\!\bigcirc\!\!-\underset{CH_3}{\overset{CH_3}{C}}-\!\!\bigcirc\!\!-O-CH_2-CH\overset{O}{-\!\!\triangle\!\!-}CH_2$$

Diglycidyl ether of bisphenol A (DGEBA)

The transformation of this viscous liquid into a hard, cross-linked three-dimensional molecular network involves the reaction of the prepolymer with reagents such as amines or Lewis acids. This reaction is referred to as curing. The curing of epoxies with a primary amine such as hexamethylenediamine involves the reaction of the amine with the epoxide. It proceeds essentially in two steps:

①The attack of an epoxide group by the primary amine

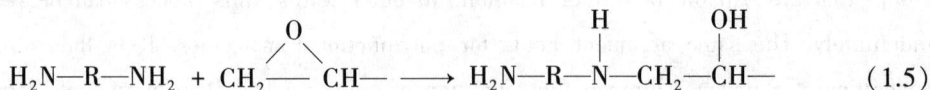

$$H_2N-R-NH_2 + CH_2\overset{O}{-\!\!\triangle\!\!-}CH- \longrightarrow H_2N-R-\overset{H}{\underset{|}{N}}-CH_2-\overset{OH}{\underset{|}{CH}}- \qquad (1.5)$$

②The combination of the resulting secondary amine with a second epoxy group to form a branch point

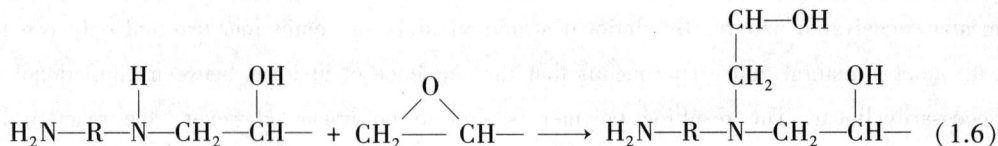

$$H_2N-R-\overset{H}{\underset{|}{N}}-CH_2-\overset{OH}{\underset{|}{CH}}- + CH_2\overset{O}{-\!\!\triangle\!\!-}CH- \longrightarrow H_2N-R-\overset{\overset{CH-OH}{\underset{|}{CH_2}}}{\underset{|}{N}}-CH_2-\overset{OH}{\underset{|}{CH}}- \qquad (1.6)$$

The presence of these branch points ultimately leads to a cross-linked infusible and insoluble polymer with structures as follows.

$$
\begin{array}{c}
\mathrm{CH-OH} \\
| \\
\mathrm{CH_2} \\
\mathrm{OH} \qquad | \qquad \mathrm{OH} \\
| \qquad\quad\;\; | \qquad\quad | \\
\mathrm{-CH-CH_2-N-R-N-CH_2-CH-} \\
| \\
\mathrm{CH_2} \\
| \\
\mathrm{CH-OH} \\
|
\end{array}
$$

In this reaction, the stoichiometric ratio requires one epoxy group per amine hydrogen. Consequently, an amine such as hexamethylenediamine has a functionality of 4. Recall, however, that in the reaction of hexamethylenediamine with adipic acid, the amine has a functionality of 2. In this reaction DGEBA is bifunctional since the hydroxyl groups generated in the reaction do not participate in the reaction. But when the curing of epoxies involves the use of a Lewis acid such as BF_3, the functionality of each epoxy group is 2; that is, the functionality of DGEBA is 4. Thus the curing reactions of epoxies further illustrate the point made earlier that the functionality of a given molecule is defined for a specific reaction. By employing different reactants or varying the stoichiometry of reactants, different structures can be produced and, consequently, the properties of the final polymer can also be varied.

Polystyrene, polyethylene, polyacrylonitrile, poly (methyl methacrylate) and poly (vinyl chloride) are typical examples of linear polymers.

Substituent groups such as $-CH_3$, $-O-\overset{\displaystyle O}{\overset{\displaystyle \|}{C}}-CH_3$, $-Cl$, and $-CN$ that are attached to the main chain of skeletal atoms are known as pendant groups. Their structure and chemical nature can confer unique properties on a polymer. For example, linear and branched polymers are usually soluble in some solvent at normal temperatures. But the presence of polar pendant groups can considerably reduce room temperature solubility. Since cross-linked polymers are chemically tied together and solubility essentially involves the separation of solute molecules by solvent molecules, cross-linked polymers do not dissolve, but can only be swelled by liquids. The presence of cross-linking confers stability on polymers. Highly cross-linked polymers are generally rigid and high-melting. Cross-links occur randomly in a cross-linked polymer. Consequently, it can be broken down into smaller molecules by random chain scission. Ladder polymers constitute a group of polymers with a regular sequence of cross-links. A ladder polymer, as the name implies, consists of two parallel linear strands of molecules with a regular sequence of cross-links. Ladder polymers have only condensed cyclic units in the chain; they are also commonly referred to as double-chain or double-strand polymers. A typical example is poly (imidazopyrrolone), which is obtained by the polymerization of aromatic dianhydrides such as pyromellitic dianhydride or aromatic tetracarboxylic acids with ortho-aromatic tetramines like 1,2,4,5-tetraaminobenzene:

9

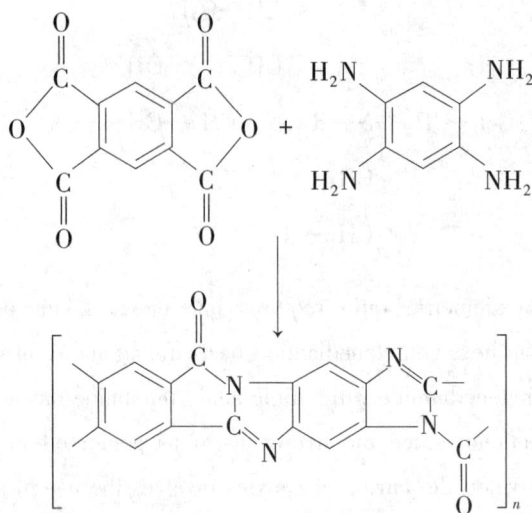

$$(1.7)$$

The molecular structure of ladder polymers is more rigid than that of conventional linear polymers. Numerous members of this family of polymers display exceptional thermal, mechanical, and electrical behavior. Their thermal stability is due to the molecular structure, which in essence requires that two bonds must be broken at a cleavage site in order to disrupt the overall integrity of the molecule; when only one bond is broken, the second holds the entire molecule together.

(2) Amorphous or crystalline

Structurally, polymers in the solid state may be amorphous or crystalline. When polymers are cooled from the molten state or concentrated from the solution, molecules are often attracted to each other and tend to aggregate as closely as possible into a solid with the least possible potential energy. For some polymers, in the process of forming a solid, individual chains are folded and packed regularly in an orderly fashion. The resulting solid is a crystalline polymer with a long-range, three-dimensional, ordered arrangement. However, since the polymer chains are very long, it is impossible for the chains to fit into a perfect arrangement equivalent to that observed in low-molecular-weight materials. A measure of imperfection always exists. The degree of crystallinity, i.e., the fraction of the total polymer in the crystalline regions, may vary from a few percentage points to about 90% depending on the crystallization conditions. Examples of crystalline polymers include polyethylene, polyacrylonitrile, poly(ethylene terephthalate), and polytetrafiuoroethylene.

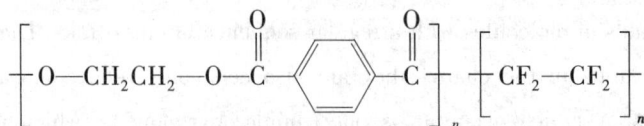

In contrast to crystallizable polymers, amorphous polymers possess chains that are incapable of ordered arrangement. They are characterized in the solid state by a short-range order of repeating units. These polymers vitrify, forming an amorphous glassy solid in which the molecular chains are arranged at random and even entangled. Poly(methyl methacrylate) and polycarbonate are typical examples.

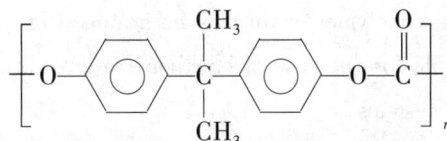

$$\left[\!\!- O \!-\!\! \bigcirc \!\!-\!\! \overset{\overset{\displaystyle CH_3}{|}}{\underset{\underset{\displaystyle CH_3}{|}}{C}} \!\!-\!\! \bigcirc \!\!-\!\! O \!-\!\! \overset{\overset{\displaystyle O}{\|}}{C} \!\!-\!\!\right]_n$$

From the above discussion, it is obvious that the solid states of crystalline and amorphous polymers are characterized by a long-range order of molecular chains and a short-range order of repeating units, respectively. On the other hand, the melting of either polymer marks the onset of disorder. There are, however, some polymers which deviate from this general scheme in that the structure of the ordered regions is more or less disturbed. These are known as liquid crystalline polymers. They have phases characterized by structures intermediate between the ordered crystalline structure and the disordered fiuid state. Solids of liquid crystalline polymers melt to form fiuids in which much of the molecular order is retained within a certain range of temperature. The ordering is sufficient to impart some solid-like properties on the fluid, but the forces of attraction between molecules are not strong enough to prevent flow. An example of a liquid crystalline polymer is polybenzamide.

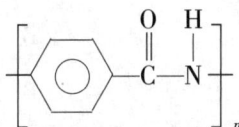

$$\left[\!\!-\!\! \bigcirc \!\!-\!\! \overset{\overset{\displaystyle O}{\|}}{C} \!-\! \overset{\overset{\displaystyle H}{|}}{N} \!\!-\!\!\right]_n$$

Liquid crystalline polymers are important in the fabrication of lightweight, ultra-high-strength, temperature-resistant fibers and films such as DuPont's Kevlar and Monsanto's X-500. The structural factors responsible for promoting the above classes of polymers will be discussed when we treat the structure of polymers.

(3) Homopolymer or copolymer

Polymers may be either homopolymers or copolymers depending on the composition. Polymers composed of only one repeating unit in the polymer molecules are known as homopolymers. However, chemists have developed techniques to build polymer chains containing more than one repeating unit. Polymers composed of two different repeating units in the polymer molecule are defined as copolymers. An example is the copolymer formed when styrene and acrylonitrile are polymerized in the same reactor. The repeating unit and the structural unit of a polymer are not necessarily the same. As indicated earlier, some polymers such as nylon 66 and poly(ethylene terephthalate) have repeating units composed of more than one structural unit. Such polymers are still considered homopolymers.

$$n\text{CH}_2\!\!=\!\!\text{CH} + m\text{CH}_2\!\!=\!\!\text{CH} \longrightarrow \left[\!\!\begin{array}{c}\text{CH}_2\!-\!\text{CH}\end{array}\!\!\right]_n\!\!\left[\!\!\begin{array}{c}\text{CH}_2\!-\!\text{CH}\\ |\\ \text{CN}\end{array}\!\!\right]_m \qquad (1.8)$$

The repeating units on the copolymer chain may be arranged in various degrees of order along the backbone; it is even possible for one type of backbone to have branches of another type. There are several types of copolymer systems:

①Random copolymer: The repeating units are arranged randomly on the chain molecule. It we represent the repeating units by A and B, then the random copolymer might have the structure shown below:

—AABBABABBAAABAABBA—

②Alternating copolymer: There is an ordered (alternating) arrangement of the two repeating units along the polymer chain:

—ABABABABABAB—

③Block copolymer: The chain consists of relatively long sequences (blocks) of each repeating unit chemically bound together:

—AAAAA—BBBBBBBB—AAAAAAAAA—BBBB—

④Graft copolymer: Sequences of one monomer (repeating unit) are "grafted" onto a backbone of the another monomer type:

```
                                    B
                                    |
                                    B
                                    |
                                    B
                                    |
                                    B
                                    |
                                    B
                                    |
                                    B
                                    |
  —AAAAAAAAAAAA—AAAAAAAA—
        |              |
        B              B
        |              |
        B              B
        |              |
        B              B
        |              |
        B              B
        |              |
        B              B
```

(4) Fibers, plastics, or elastomers

Polymers may also be classified as fibers, plastics, or elastomers. The reason for this is related to how the atoms in a molecule (large or small) are hooked together. To form bonds, atoms employ valence electrons. Consequently, the type of bond formed depends on the electronic configuration of the atoms. Depending on the extent of electron involvement, chemical bonds may be classified as either primary or secondary.

In primary valence bonding, atoms are tied together to form molecules using their valence electrons. This generally leads to strong bonds. Essentially there are three types of primary bonds: ionic, metallic, and covalent. The atoms in a polymer are mostly, although not exclusively, bonded together by covalent bonds.

Secondary bonds on the other hand, do not involve valence electrons. Whereas in the formation of a molecule atoms use up all their valence bonds, in the formation of a mass, individual molecules attract each other. The forces of attraction responsible for the cohesive aggregation between individual molecules are referred to as secondary valence forces. Examples are Van der Waals force, hydrogen, and dipole bonds. Since secondary bonds do not involve valence electrons, they are weak. (Even between secondary bonds, there are differences in the magnitude of the bond strengths: generally hydrogen and dipole bonds are much stronger than Van der Waals bonds.) Since secondary bonds are weaker than primary bonds, molecules must come together as closely as possible for secondary bonds to have maximum effect.

The ability for close alignment of molecules depends on the structure of the molecules. Those molecules with regular structure can align themselves very closely for effective utilization of the secondary intermolecular bonding forces. The result is the formation of a fiber. Fibers are linear polymers with high symmetry and high intermolecular forces that result usually from the presence of polar groups. They are characterized by high modulus, high tensile strength, and moderate extensibility (usually less than 20%). At the other end of the spectrum, there are some molecules with irregular structure, weak intermolecular attractive forces, and very flexible polymer chains. These are generally referred to as elastomers. Chain segments of elastomers can undergo high local mobility, but the gross mobility of chains is restricted, usually by the introduction of a few cross-links into the structure. In the absence of applied (tensile) stress, molecules of elastomers usually assume coiled shapes. Consequently, elastomers exhibit high elongation (up to 1 000%) from which they recover rapidly on the removal of the imposed stress. Elastomers generally have low initial modulus in tension, but when stretched they stiffen. Plastics fall between the structural extremes represented by fibers and elastomers. However, in spite of the possible differences in chemical structure, the demarcation between fibers and plastics may sometimes be blurred. Polymers such as polypropylene and polyamides can be used as fibers and as plastics by a proper choice of processing conditions.

1.3   Basic structure and properties of polymers[3-6]

Polymers, unlike organic/inorganic compounds, do not have a fixed molecular weight. It is specified in terms of degree of polymerization—number of repeat units in the chain or ration of average molecular weight of polymer to molecular weight of repeat unit. Average molecular weight is however defined in two ways. Weight average molecular weight is obtained by dividing the chains into size ranges and determining the fraction of chains having molecular weights within that range. Number average molecular weight is based on the number fraction, rather than the weight fraction, of the chains within each size range. It is always smaller than the weight average molecular weight.

Polymers are known by their high sensitivity of mechanical and/or thermal properties. This section explains their thermal behavior. During processing of polymers, they are cooled with/without presence of presence from liquid state to form final product. During cooling, an ordered solid phase may be formed having a highly random molecular structure. This process is called crystallization. The melting occurs when a polymer is heated. If the polymer during cooling retains amorphous or non-crystalline state, i.e., disordered molecular structure, rigid solid may be considered as frozen liquid resulting from glass transition. Thus, enhancement of either mechanical and/or thermal properties needs to consider crystallization, melting, and the glass transition.

Crystallization and the mechanism involved play an important role as it influences the properties of plastics. As in solidification of metals, polymer crystallization involves nucleation and growth. Near to solidification temperature at favorable places, nuclei forms, and then nuclei grow by the continued ordering and alignment of additional molecular segments. Extent of crystallization is measured by volume change as there will be a considerable change in volume during solidification of a polymer. Crystallization rate is dependent on crystallization temperature and also on the molecular weight of the polymer. Crystallization rate decreases with increasing molecular weight.

Melting of polymer involves transformation of solid polymer to viscous liquid upon heating at melting temperature, T_m. Polymer melting is distinctive from that of metals in many respects— melting takes place over a temperature range; melting behavior depends on history of the polymer; melting behavior is a function of rate of heating, where increasing rate results in an elevation of melting temperature. During melting there occurs rearrangement of the molecules from ordered state to disordered state. This is influenced by molecular chemistry and structure (degree of branching) along with chain stiffness and molecular weight.

Glass transition occurs in amorphous and semi-crystalline polymers. Upon cooling, this transformation corresponds to gradual change of liquid to rubbery material, and then rigid solid. The temperature range at which the transition from rubbery to rigid state occurs is termed as glass transition temperature, T_g. This temperature has its significance as abrupt changes in other physical

14

properties occur at this temperature. Glass transition temperature is also influenced by molecular weight, with increase of which glass transition temperature increases. Degree of cross-linking also influences the glass transition such that polymers with very high degree of cross-linking do not experience a glass transition. The glass transition temperature is typically 0.5 to 0.75 times the absolute melting temperature. Above the glass transition, non-crystalline polymers show viscous behavior, and below the glass transition they show glass-brittle behavior (as chain motion is very restricted), hence the name glass transition.

Melting involves breaking of the inter-chain bonds, so the glass- and melting- temperatures depend on:

①chain stiffness (e.g., single vs. double bonds)

②size, shape of side groups

③size of molecule

④side branches, defects

⑤cross-linking

Polymer mechanical properties can be specified with many of the same parameters that are used for metals such as modulus of elasticity, tensile/impact/fatigue strengths, etc. However, polymers are, in many respects, mechanically dissimilar to metals. To a much greater extent than either metals or ceramics, both thermal and mechanical properties of polymers show a marked dependence on parameters namely temperature, strain rate, and morphology. In addition, molecular weight and temperature relative to the glass transition play an important role that is absent for other type of materials.

A simple stress-strain curve can describe different mechanical behavior of various polymers. As shown in Figure 1. 3, the stress-strain behavior can be brittle, plastic and highly elastic (elastomeric or rubber-like). Mechanical properties of polymers change dramatically with temperature, going from glass-like brittle behavior at low temperatures to a rubber-like behavior at high temperatures. Highly crystalline polymers behave in a brittle manner, whereas amorphous polymers can exhibit plastic deformation. These phenomena are highly temperature dependent, even more so with polymers than they are with metals and ceramics. Due to unique structures of cross-linked polymers, recoverable deformations up to very high strains before point of rupture are also observed with polymers (elastomers). Tensile modulus (modulus) and tensile strengths are orders of magnitude smaller than those of metals, but elongation can be up to 1000 % in some cases. The tensile strength is defined at the fracture point and can be lower than the yield strength.

As the temperature increases, both the rigidity and the yield strength decrease, while the elongation increases. Thus, if high rigidity and toughness are the requirements, the temperature consideration is important. In general, decreasing the strain rate has the same influence on the strain-strength characteristics as increasing the temperature: the material becomes softer and more ductile. Despite the similarities in yield behavior with temperature and strain rate between polymers,

15

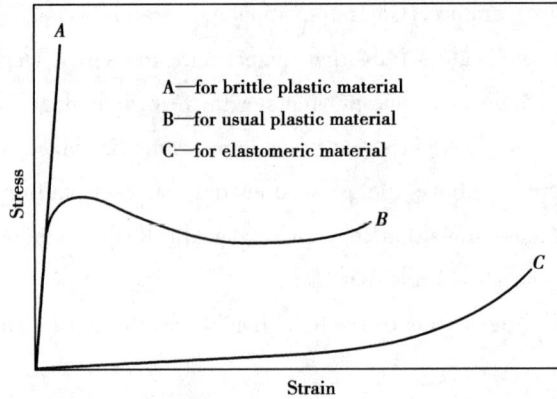

Figure 1.3 Typical stress-strain curves for polymers

metals, and ceramics, the mechanisms are quite different. Specifically, the necking of polymers is affected by two physical factors that are not significant in metals: dissipation of mechanical energy as heat, causing softening magnitude of which increases with strain rate; deformation resistance of the neck, resulting in strain-rate dependence of yield strength. The relative importance of these two factors depends on materials, specimen dimensions and strain rate. The effect of temperature relative to the glass transition is depicted in terms of decline in modulus values. Shallow decline of modulus is attributed to thermal expansion, whereas abrupt changes are attributable to viscoelastic relaxation processes.

Together molecular weight and crystallinity influence a great number of mechanical properties of polymers including hardness, fatigue resistance, elongation at neck, and even impact strength. The chance of brittle failure is reduced by raising molecular weight, which increases brittle strength, and by reducing crystallinity. As the degree of crystallinity decreases with temperature close to melting point, stiffness, hardness and yield strength decrease. These factors often set limits on the temperature at which a polymer is useful for mechanical purposes.

Elastomers, however, exhibit some unique mechanical behavior when compared to conventional plastics. The most notable characteristics are the low modulus and high deformations as elastomers exhibit large, reversible elongation under small applied stresses. Elastomers exhibit this behavior due to their unique, cross-linked structure. Elastic modulus of elastomers (resistance to the uncoiling of randomly orientated chains) increases as with increase in temperature. Unlike non-cross-linked polymers, elastomers exhibit an increase inelastic modulus with cross-link density.

An understanding of deformation mechanisms of polymers is important in order to be able to manage the optimal use of these materials, a class of materials that continues to grow in terms of use in structural applications. Despite the similarities in ductile and brittle behavior with to metals and ceramics respectively, elastic and plastic deformation mechanisms in polymers are quite different. This is mainly due to ①difference in structure they made of and ②size of the entities responsible for deformation. Plastic deformation in metals and ceramics can be described in terms of dislocations

16

and slip planes, whereas polymer chains must undergo deformation in polymers leading to different mechanism of permanent deformation. Unique to most of the polymers is the viscoelasticity—means when an external force is applied, both elastic and plastic (viscous) deformation occur. For viscoelastic materials, the rate of strain determines whether the deformation in elastic or viscous. The viscoelastic behaviour of polymeric materials is dependent on both time and temperature.

Plastic polymers deform elastically by elongation of the chain molecules from their stable conformations in the direction of the applied stress by the bending and stretching of the strong covalent bonds. In addition, there is a possibility for slight displacement of adjacent molecules, which is resisted by weak secondary / van der Waals bonds. Plastic deformation in polymers is not a consequence of dislocation movement as in metals. Instead, chains rotate, stretch, slide and disentangle under load to cause permanent deformation. This permanent deformation in polymers might occur in several stages of interaction between lamellar and intervening amorphous regions. Initial stages involve elongation of amorphous tie chains, and eventual alignment in the loading direction. Continues deformation in second stage occurs by the tilting of the lamellar blocks. Next, crystalline block segments separate before blocks and tie chains become orientated in the direction of tensile axis in final stage. This leads to highly orientated structure in deformed polymers.

Elastomers, on the other hand, deform elastically by simple uncoiling, and straightening of molecular chains that are highly twisted, kinked, and coiled in unstressed state. The driving force for elastic deformation is change in entropy, which is a measure of degree of disorder in a system. When an elastomer is stretched, the system's order increases. If elastomer is released from the applied load, its entropy increases. This entropy effect results in a rise in temperature of an elastomer when stretched. It also causes the modulus of elasticity to increase with increasing temperature, which is opposite to the behaviour of other materials.

Fracture of polymers is again dependent on morphology of a polymer. As a thumb rule, thermosets fracture in brittle mode. It involves formation of cracks at regions where there is a localized stress concentration. Covalent bonds are severed during the fracture.

However, both ductile and brittle modes are possible mode of fracture for thermoplasts. Many of thermoplasts can exhibit ductile-to-brittle transition assisted by reduction in temperature, increase in strain rate, presence of notch, increased specimen thickness and a modification of the polymer structure. Unique to polymer fracture is crazing-presence of regions of very localized yielding, which lead to formation of small and interconnected microvoids. Crazes form at highly stressed regions associated with scratches, flaws and molecular inhomogeneties; and they propagate perpendicular to the applied tensile stress and typically are 5 μm or less thick. A craze is different from a crack as it can support a load across its face.

The deformation of plastic materials can be primarily elastic, plastic, or a combination of both types. The deformation mode and resistance of deformation depends on many parameters for different plastics. The following factors influence the strength of a thermoplast: average molecular mass, degree of crystallization, presence of side groups, presence of polar and other specific atoms, presence of phenyl rings in main chains and addition of reinforcements. Effect of every one of these factors can be used to strengthen a thermoplast. Thermosets are, however, strengthened by reinforcement methods.

Strength of a thermoplast is directly dependent on its average molecular mass since polymerization up to a certain molecular-mass range is necessary to produce a stable solid. This method is not used so often as after a critical mass range, increasing the average molecular mass does not greatly increase its strength. In general, as the degree of crystallinity increases, the strength, modulus and density all increase for a thermoplast. Another method to increase the strength is to create more resistance to chain slippage. This can be achieved by addition of bulky side groups on main chains, which results in increase of strength but reduces the ductility. Increased resistance to chain slippage can be achieved by increasing the molecular bonding forces between the polymer chains. E.g., introducing a chlorine atom on every other carbon atom of main chain to make polyvinylchloride (PVC). Introducing an ether linkage (i.e., introduction of oxygen atom) or amide linkage (i.e. introduction of oxygen and nitrogen atoms) into the main chain can increase the rigidity of thermoplasts. One of the most important strengthening methods for thermoplasts is the introduction of phenylene rings in the main chain. It is commonly used for high-strength engineering plastics. The phenylene rings cause steric hindrance to rotation within the polymer chain and electronic attraction of resonating electrons between adjacent molecules. Another method of strengthening is introduction of reinforcements like glass fibers. Glass content ranges from 20% to 40%, depending on trade-off between desired strength, ease of processing and economics.

Thermosets are strengthened by reinforcements again. Different reinforcements are in use according to the necessity. Glass fibers are most commonly used to form structural and molding plastic compounds. Two most important types of glass fibers are E (electrical)- and S (high strength)- glasses. E-glass (lime-aluminium-borosilicate glass with zero or low sodium and potassium levels) is often used for continuous fibers. S-glass (65% SiO_2, 25% Al_2O_3 and 10% MgO) has higher strength-to-weight ratio and is more expansive thus primary applications include military and aerospace applications. Carbon fiber reinforced plastics are also often used in aerospace applications. However they are very expansive.

The other classes of reinforcements include aramid (aromatic polyamide) fibers. They are popularly known as Kevlar. Presently two commercial variants of Kevlar are available—Kevlar29 and Kevlar49. Kevlar29 is a low-density, high strength aramid fiber designed for applications such as ballistic protection, ropes and cables. Kevlar49 is characterized by a low density and high strength/modulus; is used in aerospace, marine, automotive and other industrial applications. Thermosets without reinforcements are strengthened by creation of network of covalent bonds throughout the structure of the material. Covalent bonds can be developed during casting or pressing under heat and pressure.

Polymers are divided into two distinct groups: thermoplastics and thermosets. The majority of polymers are thermoplastic, meaning that once the polymer is formed it can be heated and reformed over and over again. This property allows for easy processing and recycling. The other group, the thermosets, can not be remelted. Once these polymers are formed, reheating will cause the material to scorch. Every polymer has very distinct characteristics, but most polymers have the following general attributes.

①Polymers can be very resistant to chemicals. Consider all the cleaning fluids in your home that are packaged in plastic. Reading the warning labels that describe what happens when the chemical comes in contact with skin or eyes or is ingested will emphasize the chemical resistance of these materials.

②Polymers can be both thermal and electrical insulators. A walk through your house will reinforce this concept, as you consider all the appliances, cords, electrical outlets, and wiring, which are made or covered with polymeric materials. Thermal resistance is evident in the kitchen with pot and pan handles made of polymers, the coffee pot handle, the foam core of refrigerators and freezers, insulated cups, coolers, and microwave cookware. The thermal underwear that many skiers wear is made of polypropylene and the fiberfill in winter jackets is acrylic.

③Generally, polymers are very light in mass with varying degrees of strength. Consider the range of applications, from dime store toys to the frame structure of space stations, or from delicate nylon fiber in pantyhose to Kevlar^{TM}, which is used in bulletproof vests.

④Polymers can be processed in various ways to produce thin fibers or very intricate parts. Plastics can be molded into bottles or the body of a car or be mixed with solvents to become an adhesive or a paint. Elastomers and some plastics stretch and are very flexible. Other polymers can be foamed like polystyrene (Styrofoam^{TM}) and urethane, to name just two examples. Polymers are materials with a seemingly limitless range of characteristics and colors. Polymers have many inherent properties that can be further enhanced by a wide range of additives to broaden their uses and applications.

1.4   Applications of polymers[6-10]

Polymers are increasingly being used as a substitute for conventional material systems and are combined with conventional materials to produce high quality hybrid systems. Polymer science has had a major impact on the way we live. It is difficult to find an aspect of our lives that is not affected by polymers. Just 100 years ago, materials we now take for granted were non-existent. With further advances in the understanding of polymers, and with new applications being researched, there is no reason to believe that the revolution will stop any time soon. This section presents some common applications of a few polymers that will be detailedly introduced in the textbook.

1.4.1　Thermo plastics

(1) Acrylonitrile-butadiene-styrene (ABS)

Characteristics: Outstanding strength and toughness, resistance to heat distortion; good electrical properties; flammable and soluble in some organic solvents.

Application: Refrigerator lining, lawn and garden equipment, toys, highway safety devices.

(2) Acrylics (poly-methyl-methacrylate)

Characteristics: Outstanding light transmission and resistance to weathering; only fair mechanical properties.

Application: Lenses, transparent aircraft enclosures, drafting equipment, outdoor signs.

(3) Fluorocarbons (PTFE or TFE)

Characteristics: Chemically inert in almost all environments, excellent electrical properties; low coefficient of friction; may be used to 260 ℃; relatively weak and poor cold-flow properties.

Application: Anticorrosive seals, chemical pipes and valves, bearings, anti adhesive coatings, high temperature electronic parts.

(4) Polyamides (nylons)

Characteristics: Good mechanical strength, abrasion resistance, and toughness; low coefficient of friction; absorbs water and some other liquids.

Application: Bearings, gears, cams, bushings, handles, and jacketing for wires and cables.

(5) Polycarbonates

Characteristics: Dimensionally stable: low water absorption; transparent; very good impact resistance and ductility.

Application: Safety helmets, lenses light globes, base for photographic film.

(6) Polyethylene

Characteristics: Chemically resistant and electrically insulating; tough and relatively low coefficient of friction; low strength and poor resistance to weathering.

Application: Flexible bottles, toys, tumblers, battery parts, ice trays, film wrapping materials.

(7) Polypropylene

Characteristics: Resistant to heat distortion; excellent electrical properties and fatigue strength; chemically inert; relatively inexpensive; poor resistance to UV light. Application: Sterilizable bottles, packaging film, TV cabinets, luggage.

(8) Polystyrene

Characteristics: Excellent electrical properties and optical clarity; good thermal and dimensional stability; relatively inexpensive.

Application: Wall tile, battery cases, toys, indoor lighting panels, appliance housings.

(9) Polyester (PET or PETE)

Characteristics: One of the toughest of plastic films; excellent fatigue and tear strength, and resistance to humidity acids, greases, oils and solvents.

Application: Magnetic recording tapes, clothing, automotive tire cords, beverage containers.

1.4.2   Thermo setting polymers

(1) Epoxies

Characteristics: Excellent combination of mechanical properties and corrosion resistance; dimensionally stable; good adhesion; relatively inexpensive; good electrical properties.

Application: Electrical moldings, sinks, adhesives, protective coatings, used with fiberglass laminates.

(2) Phenolics

Characteristics: Excellent thermal stability to over 150 ℃; may be compounded with a large number of resins, fillers, etc.; inexpensive.

Application: Motor housing, telephones, auto distributors, electrical fixtures.

Collection of Exercises

1. Many polymers occur in nature. Name any two naturally occurring polymers.
2. What will happen when we begin heating each of the polymers below?
a. A thermosetting (polymer).
b. A thermoplastic (polymer).
3. Name any two properties that you may look for in describing a polymer.
4. Describe what is meant by the term "polymer".
5. Briefly explain the difference in molecular chemistry between silicone polymer and other polymeric materials.

REFERENCES

[1] Paul C. Hiemenz. Polymer Chemistry: The Basic Concepts, Marcel Dekker, Inc. New York, 1984.

[2] Robert O. Ebewele. Polymer science and technology, CRC Press LLC, Boca Raton, 2000.

[3] V. R. Gowariker, N. V. Viswanathan, and Jayadev Sreedhar, Polymer Science, New Age

International Private Limited publishers, Bangalore, 2001.

[4] C. A. Harper, Handbook of Plastics Elastomers and Composites, Third Edition, McGrawHill Professional Book Group, New York, 1996.

[5] William D. Callister, Jr, Materials Science and Engineering-An introduction, 5th edition, John Wiley & Sons, Inc. New York, 2004.

[6] R.J. Young , Introduction to Polymers, Chapman and Hall, London, 1981.

[7] K. Hatada, Eds. T. Kitayama, Macromolecular Design of Polymeric Materials, Marcel Dekker, New York, 1999.

[8] Christopher Hall, Polymer Materials—An introduction for technologists and scientists, John Wiley & Sons, Inc., New York, 1981.

[9] H. Ulrich, Introduction to Industrial Polymers, Henser Publishers, Munich, 1989.

[10] Jr., F. W. Billmeyer, Text Book of polymer Science, 3rd ed., Wiley Intersciene, New York, 1984.

CHAPTER 2

COMMODITY PLASTICS

2.1 Introduction

Plastics is a kind of synthetic polymer materials which has been used widely, plastic products can be found everywhere in our daily life. From the toiletries we used after getting up, tableware, breakfast to stationery while we work or study with, cushions, mattresses during the break, and TV sets, washing machines, computer shell, as well as all sorts of lamps and lanterns which bring us light at night, etc.

Due to its superior performance, plastics gradually take the place of many materials and utensils which have been used for decades or hundreds of years, plastics become an indispensable assistant in people's life. Plastics has many advantages, such as the stiffness of the metal, the portability of lumber, the transparency of glass, the corrosion resistance pottery and porcelain of, the elasticity and toughness of rubber, so in addition to daily necessities, plastics are more widely used in aerospace, medical equipment, petrochemical industry, machinery manufacturing, defense, construction and so on all walks of life.

2.2 Properties and classifications of plastics

There are many kinds of plastics, until now there are about more than three hundred kinds of plastics. The basic structure of plastics (or polymers) is given by macromolecule chains, formulated from monomer units by chemical reactions. Typical reactions for chain assembling are polyaddition (continuous or stepwise) and condensation polymerization (polycondensation)[1] (Figure 2.1).

①Polyaddition as chain reaction: Process by chemical combination of a large number of

23

monomer molecules, in which the monomers will be combined to a chain either by orientation of the double bond or by ring splitting. No byproducts will be separated and no hydrogen atoms will be moved within the chain during the reaction. The process will be started by energy consumption (by light, heat or radiation) or by use of catalysts.

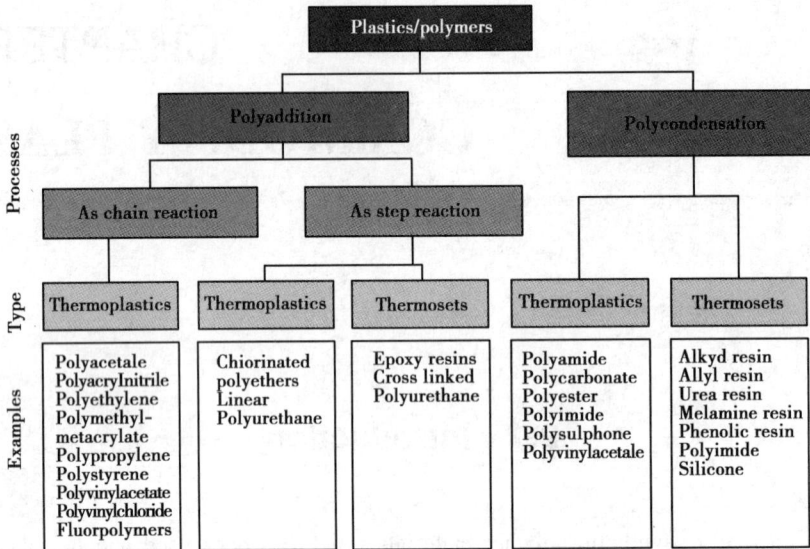

Figure 2.1　Processes for generating plastics and examples

②Polyaddition as step reaction: Process by combination of monomer units without a reaction of double bonds or separation of low molecular compounds. Hydrogen atoms can change position during the process.

③Polycondensation: Generation of plastics by build up of polyfunctional compounds. Typical small molecules like water or ammonia can be set free during the reaction. The reaction can occur as a step reaction.

The monomer units are organic carbon-based molecules. Beside carbon and hydrogen atoms as main components elements like oxygen, nitrogen, sulfur, fluorine or chlorine can be contained in the monomer unit. The type of elements, their proportion and placing in the monomer molecule gives the basis for generating different plastics, as shown in Table 2.1.

Table 2.1　Examples of some common plastics and their monomers

Monomer		Polymer	
Ethylene	$CH_2{=}CH_2$	Polyethylene(PE)	$\left[CH_2{-}CH_2\right]_n$
Propylene	$\underset{\underset{CH_3}{\vert}}{CH{=}CH_2}$	Polypropylene(PP)	$\underset{\underset{CH_3}{\vert}}{\left[CH{-}CH_2\right]_n}$
Vinylchloride	$CH_2{-}C\overset{H}{\underset{Cl}{\diagdown}}$	Polyvinylchloride(PVC)	$\underset{\underset{Cl}{\vert}}{\left[CH{-}CH_2\right]_n}$

continued

Monomer		Polymer	

| Caprolactame | CH_2 — CH_2 — $N-H$ — $C=O$ — CH_2 — CH_2 — CH_2 | Poly(E-Caprolactame) (PA6) | $\begin{array}{c} O \\ \parallel \\ -[NH-(CH_2)_5-C]_n- \end{array}$ |
| Tetraflourethylene $CF_2{=}CF_2$ | Polytetraflourethylene (PTFE) | $-[CF_2-CF_2]-$ | $-[CF_2-CF_2]_n-$ |

The coupling between the atoms of a macromolecular chain happens by primary valence bonding. The backbone of the chain is built by carbon atoms linked together by single or double bonding. Given by the electron configuration of carbon atoms, the link between the carbon atoms occurs at a certain angle, for example, for single bonding at an angle of 109.5°. Atoms like hydrogen, which are linked to the carbon atoms, hinder the free rotation of the carbon atoms around the linking axis.

The "cis"-link of carbon atoms has the highest bonding energy while the "trans"-link has the lowest (Figure 2.2). Depending on the type of bonding partners several chain conformations are possible. Examples of such conformations are zig-zag conformation (e.g., PE or PVC) or helix conformation (e.g., PP, POM or PTFE) (Figure 2.3).

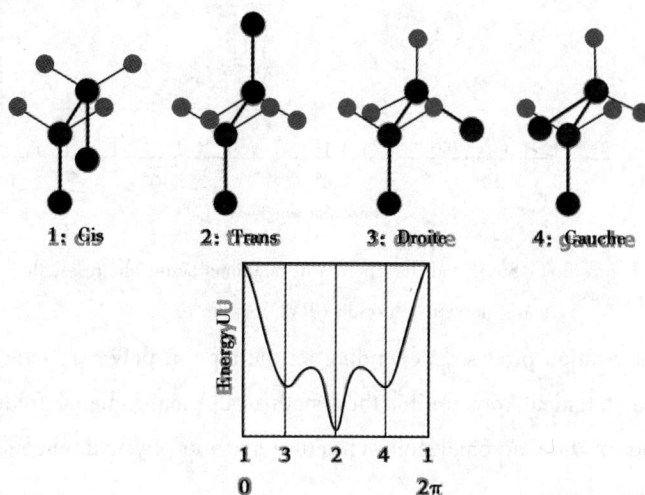

1: Cis   2: Trans   3: Droite   4: Gauche

Figure 2.2   Potential energy for rotation of
ethylene molecules around the carbon-linking axis

The chain length and by this also the molecular weight of macromolecules have a statistical distribution (Figure 2.4). By influencing the conditions of the polymerization process, the average molecular weight and the width of the distribution function can be controlled within certain limits.

(a)Linear conformation

(b)Zig-zag conformation

(c)Helix-conformation

(d)Ball-shaped conformation

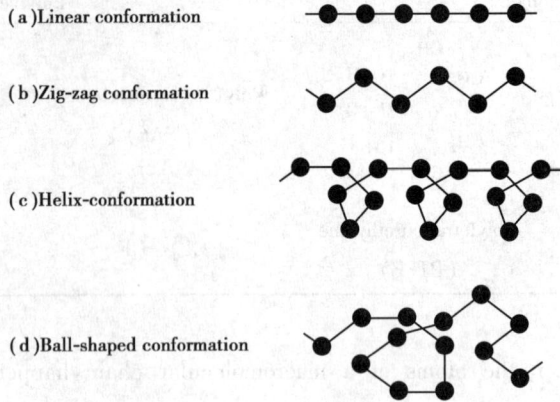

Figure 2.3   Conformation types of macromolecules

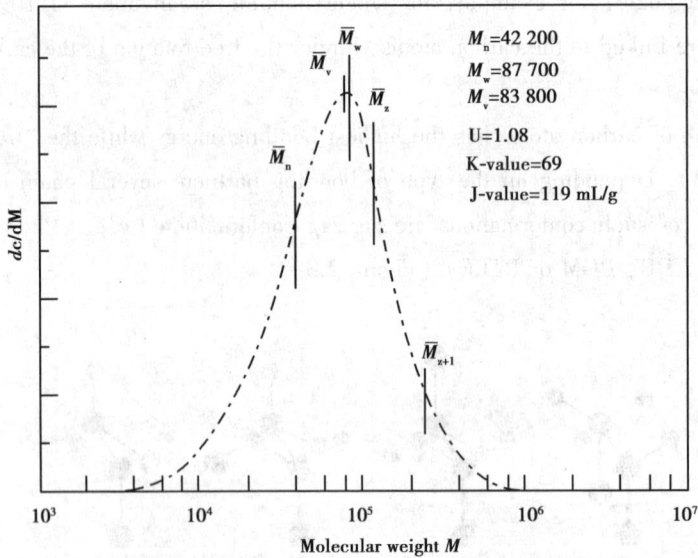

$\overline{M}_w$

$\overline{M}_v$

$\overline{M}_z$

$\overline{M}_n$

M_n=42 200
M_w=87 700
M_v=83 800

U=1.08
K-value=69
J-value=119 mL/g

dc/dM

$\overline{M}_{z+1}$

10^3 10^4 10^5 10^6 10^7

Molecular weight M

Figure 2.4   Statistical distribution of macromolecule chain length
using polyvinylchloride (PVC) as an example

During the polymerization process, depending on the type of polymer, side chains can be built to the main chain in a statistical way. As for the length of the main chain, frequency and length of the side chains depend on the macromolecular structure and the physical/chemical conditions of the polymerization process.

An example for the order of size of macromolecules is the length and width of polystyrene molecules with an average molecular weight of 10^5. Corresponding to the molecular weight the macromolecular chain consists of a number of approximately 2×10^5 carbon atoms. The average distance between each carbon atom is 1.26×10^{-10} m. Using this distance and the number of atoms in the chain takes to a length of 25×10^{-6} m and $(4 \sim 6) \times 10^{-10}$ m width for a stretched chain.

The statistical forming of the macromolecular structure of plastics results in the fact that physical properties of plastics, like temperatures of phase changes, can only be given as average values. Unlike materials like metals, phase changes of plastics occur in certain temperature ranges. The width of such temperature ranges is dependent on the homogeneity of the materials structure.

The physical and chemical structure of the macromolecule is given by the primary valence bonding forces between the atoms (Figure 2.5). The secondary valence bonding forces, like dispersion bonding, dipole bonding or hydrogen bridge bonds, have a direct influence to the macroscopic properties of the plastic like mechanical, thermal, optical, electrical or chemical properties.

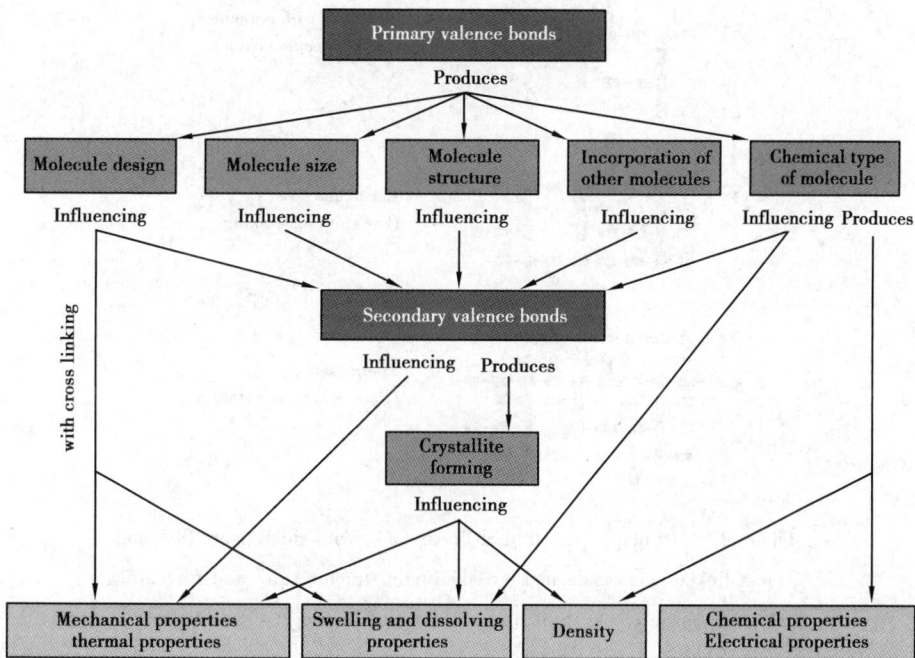

Figure 2.5   Context of molecular and macroscopic material properties

The secondary valence forces are responsible for the orientation of the macro- molecules among themselves. During processing of plastics the orientation of molecule segments can result in an orientation of segments of the macromolecular chain. Under suitable conditions, like specific placements of atoms in the monomer structure and by this within the macromolecular chain, a partial crystallization of the plastic is possible. The strength of the secondary valences is directly correlated with the formation of the macromolecular chains. The strength increases with increasing crystallization, with higher polarity between the monomer units, decreased mobility of molecule segments and increased strapping of chains with others. Because of the small range and low energy of secondary valences in comparison with the main valences, effects caused by them are strongly temperature dependent.

In the case of possible atom bonds between macromolecular chains, a crosslinking of the molecule structure can happen. While secondary valences can be dissolved with increasing

27

temperatures and rebuilt during cooling, atom bonds cannot dissolve reversibly. By dissolving these bonds the plastic will be chemically destroyed.

Taking the chemical structure and the degree of crosslinking between the macromolecules, plastics can be classified as thermoplastics, elastomers and thermosets (Figure 2.6). Compounds like polymer blends, copolymers and composite materials are composed of several base materials. This composition can be done on a physical basis (e.g., polymer blends or composite materials) or on a chemical basis (copolymers).

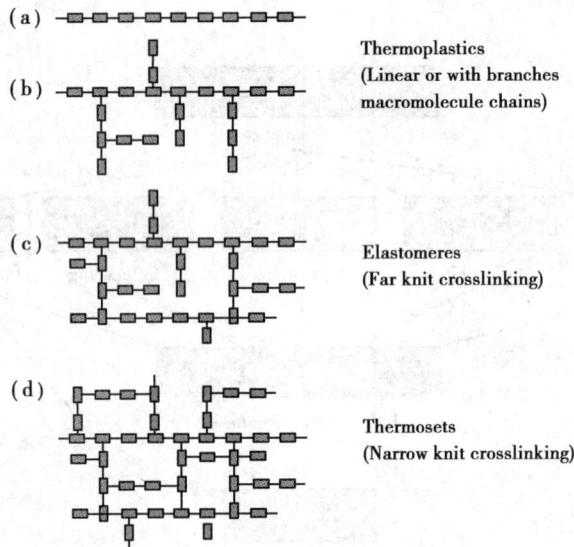

(a)

Thermoplastics
(Linear or with branches
macromolecule chains)

(b)

(c)

Elastomeres
(Far knit crosslinking)

(d)

Thermosets
(Narrow knit crosslinking)

Figure 2.6 Principle structure of linear (a), with side chains (b) and
cross-linked macromolecules (c+d).Chain structure (a) and (b) are
thermoplastic types,structures with low crosslinking (c) elastomers
and with strong crosslinking thermosets (d)

Caused by the macromolecular structure and the temperature-dependent physical properties plastic materials are distinguished into different classes. Figure 2.7 gives an overview of the classification of plastics with some typical examples.

Thermoplastics are in the application range of hard or tough elasticity and can be melted by energy input (mechanical, thermal or radiation energy). Elastomers are of soft elasticity and usually cannot be melted. Thermosets are in the application range of hard elasticity and also cannot be melted.

Plastics as polymer mixtures are composed of two or more polymers with homogeneous or heterogeneous structure. Homogeneous structures are for example copolymers or thermoplastic elastomers, built by chemical composition of two or more different monomer units in macromolecules. When using thermoplastic monomers such plastic material can be melted by thermal processes. Heterogeneous structures are for example polymer blends or thermoplastic elastomers, built by physical composition of separate phases from different polymers. Polymer blends with thermoplastic components also can be melted by thermal processes.

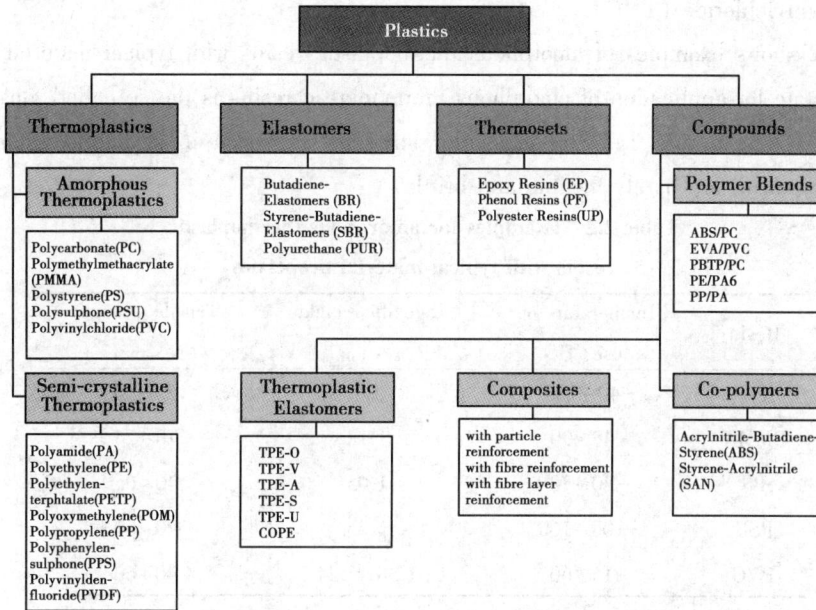

Figure 2.7   Classification of plastics

Plastic composites consist of a polymeric matrix with integrated particles or fibers. When using thermoplastics as matrix, such composites can be melted. If thermosets are used as matrix the composite cannot be melted.

Characteristic of the different classes of plastics are the phase transitions that occur in contrast to metallic materials in temperature intervals. Phase-transition temperatures are dependent on the molecular structure of the plastic. Limited mobility of the molecule chains, for example, by loop forming, long side chains or high molecular weight cause an increased phase-transition temperature. A large variance of the molecule chain length or number and length of side chains also have an effect on the spreading of the phase-transition ranges.

Thermoplastic resins consist of macromolecular chains with no crosslinks between the chains. The macromolecular chains themselves can have statistical oriented side chains or can build statistical distributed crystalline phases. The chemistry and structure of thermoplastic resins have an influence on the chemical resistance and resistance against environmental effects like UV radiation. Naturally, thermoplastic resins can vary from optical transparency to opaque, depending on the type and structure of the material. In opaque material, the light is internally scattered by the molecular structure and direct transmission of light is very poor with increasing material thickness.

Thermoplastic resins can be reversibly melted by heating and re-solidified by cooling without significant changing of mechanical and optical properties. Thus, typical industrial processes for part manufacturing are extrusion of films, sheets and profiles or molding of components.

Amorphous thermoplastic resins consist of statistical oriented macromolecules without any near order. Such resins are in general optically transparent and mostly brittle. Typical amorphous thermoplastic resins are polycarbonate (PC), polymethyl-methacrylate (PMMA), polystyrene (PS) or polyvinylchloride (PVC).

Table 2.2 shows examples of amorphous thermoplastic resins with typical material properties. Temperature state for application of amorphous thermoplastic resins is the so called glass condition below the glass temperature Tg. The molecular structure is frozen in a definite shape and the mechanical properties are barely flexible and brittle.

Table 2.2 Examples for amorphous thermoplastic resins with typical material properties

Resin	Temperature of use/℃	Specific weight /($g \cdot cm^{-3}$)	Tensile strength /($N \cdot mm^{-2}$)
PC	−40~120	1.2	65~70
PMMA	−40~90	1.18	70~76
PS	−20~70	1.05	40~65
PSU	−100~160	1.25	70~80
PVC	−15~60	1.38~1.24	40~60

Semicrystalline thermoplastic resins consist of statistical oriented macromolecule chains as amorphous phase with embedded crystalline phases, built by near-order forces. Such resins are usually opaque and tough elastic. Typical semicrystalline thermoplastic resins are polyamide (PA), polypropylene (PP) or high density polyethylene (HDPE) (Table 2.3).

Table 2.3 Examples for semicrystalline thermoplastic resins with typical material properties

Resin	Temperature of use/℃	Crystallization grade/%	Specific weight /($g \cdot cm^{-3}$)	Tensile strength /($N \cdot mm^{-2}$)
PA6	−40~100	20~45	1.12~1.15	38~70
HDPE	−50~90	65~80	0.95~0.97	19~39
PETP	−40~110	0~40	1.33~1.38	37~80
PP	−5~100	55~70	0.90~0.91	21~37
PPS	<230	30~60	1.35	65~85
PVDF	−30~150	52	1.77	30~50

Thermosets are plastic resins with narrow cross-linked molecule chains[1]. Examples of thermosets are epoxy resin (EP), phenolic resin (PF) or polyester resin (UP).

In the state of application, thermosets are hard and brittle. Because of the strong resistance of molecular movement caused by the crosslinking, mechanical strength and elasticity are not temperature dependent, as with thermoplastics or elastomers.

Thermosets cannot be melted and joining by thermal processes like ultrasonic welding or laser welding is not possible. On exceeding the decomposition temperature T_d, the material will be chemical decomposed.

2.3   Plastic-processing methods and design guidelines[2]

2.3.1   Introduction

Both natural and synthetic polymers must be processed before use. The seeds must be separated from cotton in the ginning process, pigments and driers must be added to oleoresinous paints, and the latex of Hevea rubber or gutta-percha must be coagulated to obtain the solid elastomer plastic. Synthetic polymers must also be compounded and fabricated into useful shapes. Plastics are converted into their final shapes by utilizing a variety of techniques and machinery.

Polymer processing can be defined as the process whereby raw materials are converted into products of desired shape and properties. Thermoplastic resins are generally supplied as pellets, marbles, or chips of varying sizes, and they may contain some or all of the desired additives. When heated above their T_g, thermoplastic materials soften and flow as viscous liquids that can be shaped using a variety of techniques and then cooled to "lock" in the micro- and gross structure.

Thermosetting feedstocks are normally supplied as meltable and/or flowable prepolymer, oligomers, or lightly or non-cross-linked polymers that are subsequently cross-linked, forming the thermoset article.

The processing operation can be divided into three general steps: pre-shaping, shaping, and post-shaping. In pre-shaping, the intent is to produce a material that can be shaped by application of heat and/or pressure. Important considerations include:

①Handling of solids and liquids including mixing, low, compaction, and packing

②Softening through application of heat and/or pressure

③Addition and mixing/dispersion of added materials

④Movement of the resin to the shaping apparatus through application of heat and/or pressure and other flow aiding processes

⑤Removal (and desired and recycling) of unwanted solvent, unreacted monomer(s), byproducts, and waste (flash)

The shaping step may include any single or combination of the following:

①Die forming (including sheet and film formation, tube and pipe formation, fiber formation, coating, and extrusion)

②Molding and casting

③Secondary shaping (such as film and blow molding, thermoforming)

④Surface treatments (coating and calendering)

Post-shaping processes include welding, bonding, fastening, decorating, cutting, milling, drilling, dying, and gluing.

Polymer processing operations can be divided into five broad categories:

①Spinning (generally for fibers)

②Calendering

③Coating

④Molding

⑤Injection

Processing, chemical structure, physical structure (amorphous/crystalline), and performance are interrelated to one another. Understanding these factors and their interrelationships becomes increasingly important as the specific performance requirements become more specific. Performance is related to the chemical and physical structure and to the particular processing performed on the material during its lifetime. The physical structure is a reflection of both the chemical structure and the total history of the synthesis and subsequent exposure of the material to additional force that contributes to the secondary (and greater) structure-stress/strain, light, chemical, and so on.

A single material may be processed using only a single process somewhat unique to that material (such as liquid crystals) or by a variety of processes (such as polyethylene) where the particular technique is dictated by such factors as end us and cost.

Following is a brief description of some of the most widely used techniques employed in the processing of plastics.

2.3.2 Casting

One of the simplest and least expensive methods for the production of plastic articles is casting. In this process, which is illustrated in Figure 2.8, a prepolymer, such as a catalyzed epoxy resin, is placed in a mold and allowed to harden, preferably with additional heat. This technique may also be used with urethane reactants (RIM), phenolic resins, unsaturated polyesters, PVC plastisols, and acrylic resins.

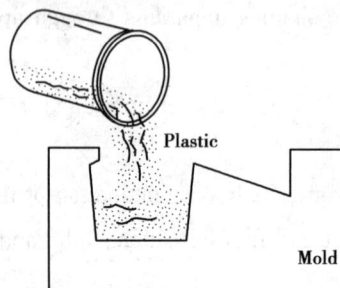

Figure 2.8   Illustration of the casting method
of molding plastics

With the exception of plastisols, most of these processes are exothermic and thus the articles should be small or the mold must be cooled. Plastisols, which consist of a dispersion of a finely divided polymer, usually PVC, in a liquid plasticizer, must be heated to at least 150 ℃ to fuse the plasticizer-polymer mixture. Polymers, like ethylcellulose and ethylene-vinyl acetate copolymers, which can be melted without decomposition, can be cast as hot melts. Solutions of polymers can be cast as films.

Polymer concrete is produced by a casting process. Simulated marble consists of a filled-peroxy-catalyzed unsaturated polyester prepolymer that polymerizes in situ. Comparable mortars consisting of filled-catalyzed phenolic, epoxy, or polyester resins are used for joining brick and tile. Casting is used in manufacturing both thermosetting and thermoplastic resins for making eyeglass lenses, plastic jewelry, and cutlery handles.

2.3.3 Blow molding

Blow molding and plug-assisted vacuum thermoforming are employed to make hollow items such as bottles and many hollow, thin-walled toys and bowls. For blow molding, a plastic parison is placed in the mold and air is applied through the opening of the cylinder-shaped plastic, blowing the plastic toward the mold walls (Figure 2.9). In the plug-assisted molding sequence the plastic resin is present as a sheet.

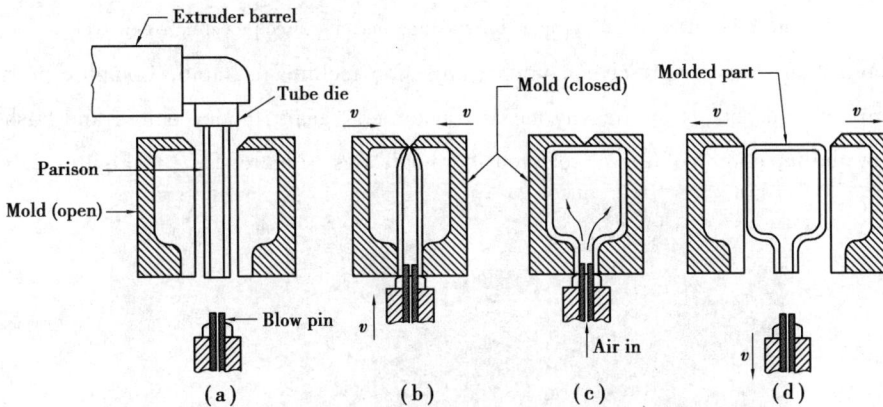

Figure 2.9　Blow molding technique

2.3.4 Injection molding

In injection molding, a large volume of thermoplastics is injection-molded to produce a variety of articles at a rapid rate. As shown in Figure 2.10, the polymer pellets may be heated, softened, and formed or forced (injected) by a screw into a closed, cooled mold. The split mold is opened and closed after the molded article is ejected and the cycle is then repeated. As shown in Figure 2.11, a reciprocating preplasticating screw moves forward to eject the softened polymer.

Figure 2.10   A large (3 000 ton capacity) injection molding
machine (courtesy Cincinnati Milacron)

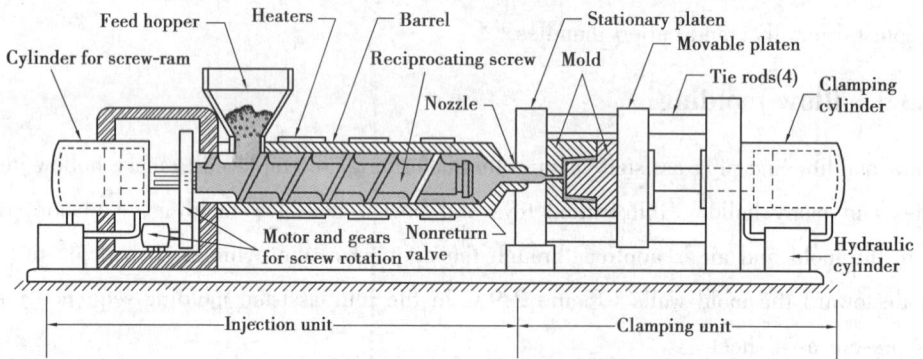

Figure 2.11   Diagram of an injection molding machine, reciprocating screw type

In contrast to slow compressive molding, injection molding is rapid. Complex parts may be produced in a few seconds in multicavity molds. Containers, gears, honeycombs, and trash cans are produced by the injection molding of selected thermoplastics (Figures 2.12 to Figures 2.16) .

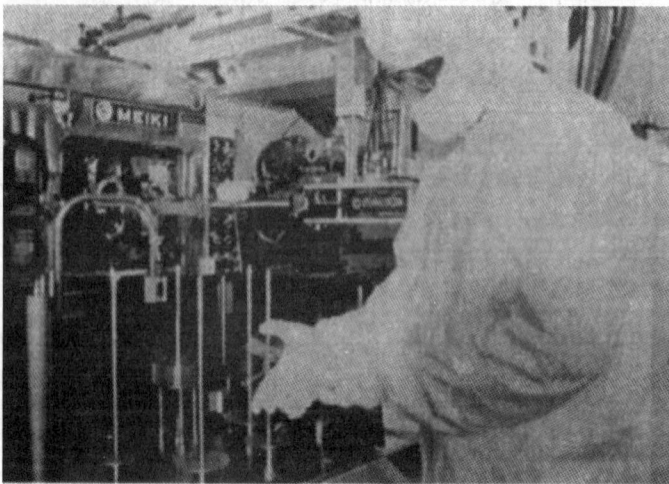

Figure 2.12   A Mobay laboratory technician inspects a compact disc that was
molded on the Meiki injection-molding system shown in the background.

Figure 2.13   A Sailor robot automatically removes a
compact disc from the mold.

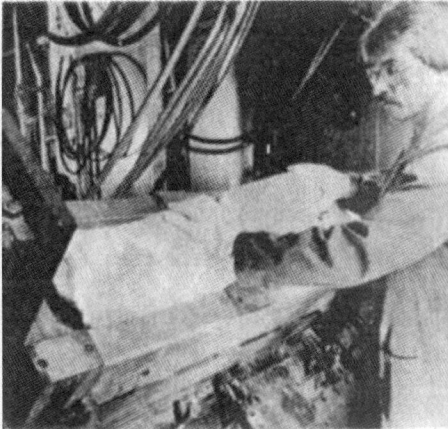

Figure 2.14   A Mobay lab technician places
combinations of preformed glass reinforcement
for a bumper beam in the RIM tool.

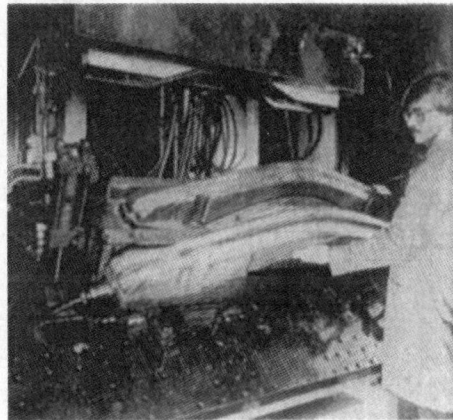

Figure 2.15   A finished bumper beam is removed
from the mold less than a minute after the
polyurethane mixture is injected into the tool.

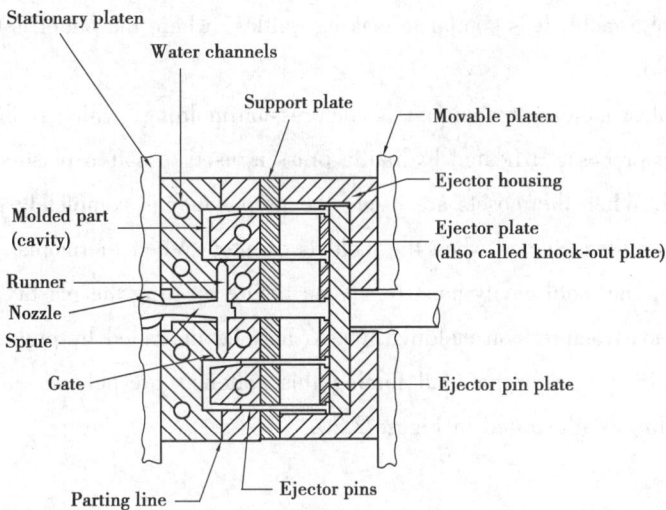

Stationary platen
Water channels
Support plate
Movable platen
Ejector housing
Molded part
(cavity)
Ejector plate
(also called knock-out plate)
Runner
Nozzle
Sprue
Gate
Ejector pin plate
Parting line
Ejector pins

Figure 2.16   Structure of injection mold

2.3.5 **Laminating**

In laminating, sheets of metal foil, paper, other plastic, or cloth are treated with a plastic resin. They are then run through rollers that squeeze the sheets together and heat them as shown in Figure 2.17. Paneling and electronic circuits are examples of products produced through this process, which is similar to making sandwiches.

Calendering is similar to laminating except rollers spread melted resin over the sheets to be covered, providing a protective coating as in the case of playing cards and treated wallpapers (Figure 2.18). This is similar to spreading jelly on a slice of bread, with the jelly being the resin.

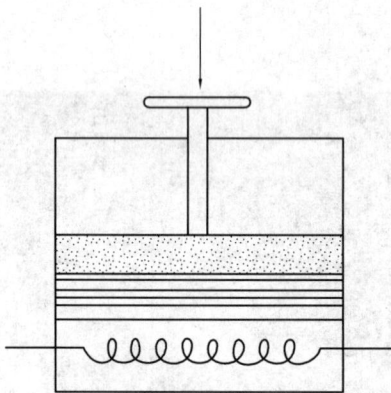

Figure 2.17   Assembly illustrating the
laminating process

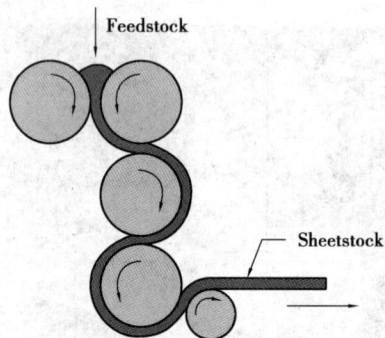

Figure 2.18   A typical roll configuration
in calendering

2.3.6 **Compression molding**

There are a wide variety of molding techniques. Simple molding entails squeezing plastic between two halves of a mold. It is similar to making waffles, where the batter is the plastic and the waffle iron is the mold.

One of the simplest molding techniques is compression molding, which is illustrated in Figure 2.19. In this molding process, a heated hydraulic press is used to soften plastic pellets and shape the plastic in a mold. When thermosets are used, the prepolymer is completely polymerized in the closed hot mold and is then ejected when the mold is opened. When thermoplastics are molded by compression molding, the mold cavity must be cooled before ejecting the plastic article.

The labor-intensive compression molding process may be upgraded by preheating a preform of the molding powder in a transfer pot and forcing this softened pre-polymer into hot multicavity molds. Transfer molding is illustrated in Figure 2.20.

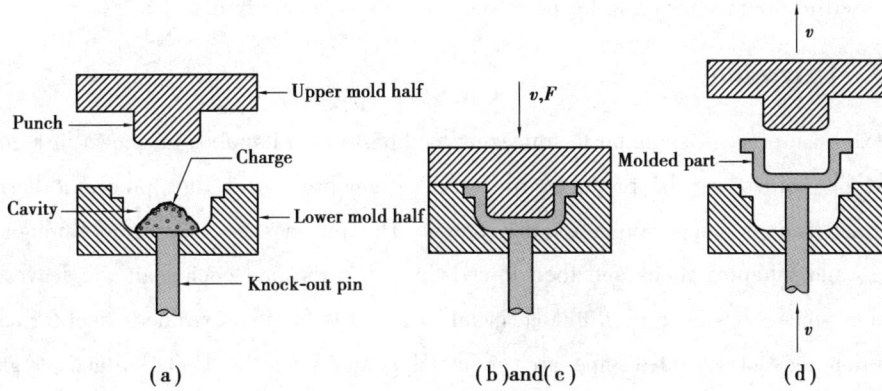

Figure 2.19   Compression molding technique

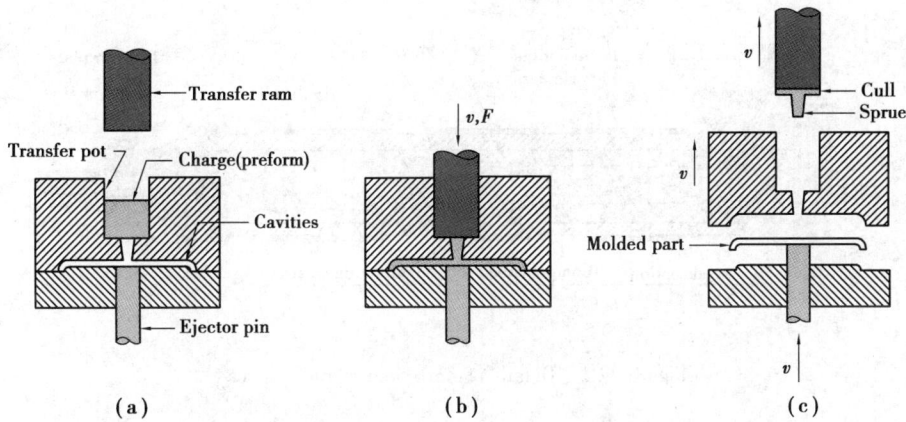

Figure 2.20   Pot transfer molding (a) charge is loaded into pot, (b) softened
polymer is pressed into mold cavity and cured, and (c) part is ejected.

2.3.7   Rotational molding

One of the more versatile molding techniques is rotational molding in which a hollow mold containing a resin powder or a liquid plastisol is heated and rotated simultaneously on two perpendicular axes. The mold is then cooled and the hollow object, such as a pipe fitting, is removed.

2.3.8   Calendering

One of the most commonly used techniques for making thermoplastic or elastomeric sheet is calendering. Calendering is a method of producing plastic film and sheet by squeezing the plastic through the gap between two counter-rotating cylinders. The art of forming a sheet in this way can be traced to the paper, textile and metal industries[3]. As shown in Figure 2.18, the polymer is transported through heated rollers, like those in a rubber mill, to a series of heated wringer-type rollers, which press the polymer into a continuous sheet of uniform thickness. The calendering process is used to produce sheeting for upholstery and for thermoforming.

2.3.9 Extrusion

The extrusion process is similar to squeezing toothpaste from its tube. As shown in Figure 2.21, pipe, rods, or profiles may be produced by the extrusion process. In this process, thermoplastic pellets are fed from a hopper to a rotating screw. The polymer is transported through heated, compacting, and softening zones and then forced through a die and cooled after it leaves the die. The extrusion process has been used to coat metal wire and to form co-extruded sheet for packaging. Over 1 million tons of extruded pipe are produced annually in the United States. Figure 2.22 illustrates film formation employing the extrusion process, and Figures 2.23 and 2.24 show an automated extrusion production line.

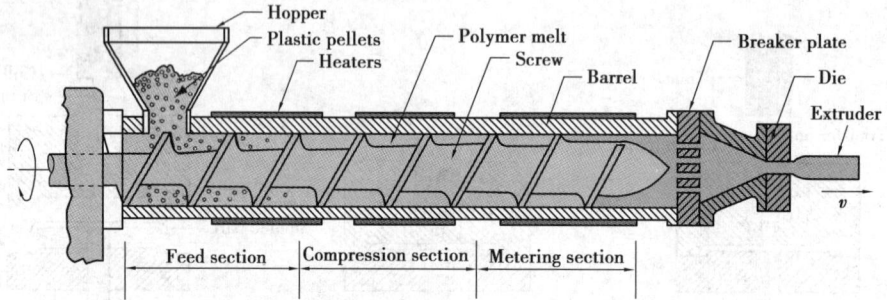

Figure 2.21   Details of screw and extruder zones

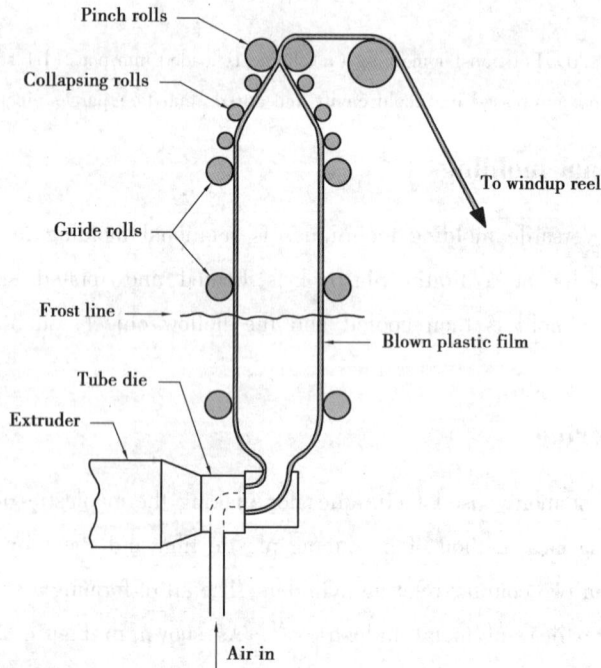

Figure 2.22   Diagram of film formation employing the extrusion process

Figure 2.23   The entire extrusion line in Mobay's
laboratories is controlled by state-of-the-art
microprocessor technology, which was designed
to optimize processing parameters and reduce
start-up times.

Figure 2.24   An overview of the new Mobay
multipurpose extrusion line

2.3.10   Thermoforming

Thermoplastic sheet, produced by extrusion through a slit die, calendering, or hot pressing of several calendered sheets, is readily thermoformed by draping over a mold and using a plunger or vacuum to force the sheet into the shape of the mold. As illustrated in Figure 2.25, refrigerator liners or suitcases may be produced by vacuum sheet thermoforming.

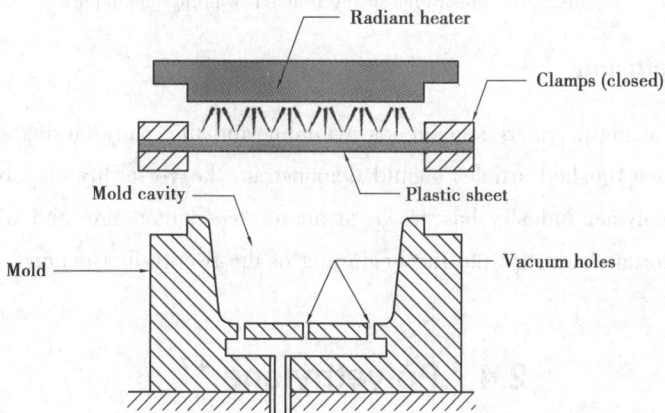

Figure 2.25   Vacuum thermoforming.

2.3.11　Reinforced plastics

Fiberglass-reinforced plastics (FRP) are fabricated by casting procedures using mixtures of the casting resins and glass or graphite fibers. In the simplest hand lay-up technique, a catalyzed resin, such as an unsaturated polyester resin, is placed on a male form. This gel coat formulation is followed by a sequential buildup of layers of catalyzed resin-impregnated glass mat. The composite is removed after it hardens. The curing step may be accelerated by heating. This technique may be modified by spraying a mixture of chopped fibers and the catalyzed prepolymer onto the form.

In a more sophisticated approach, a continuous resin-impregnated filament is wound around a rotating mandrel and cured as shown in Figure 2.26. In another modification, a bundle of resin-impregnated filaments is drawn through a heated die. Fishing rods and pipe are produced by this pultrusion technique.

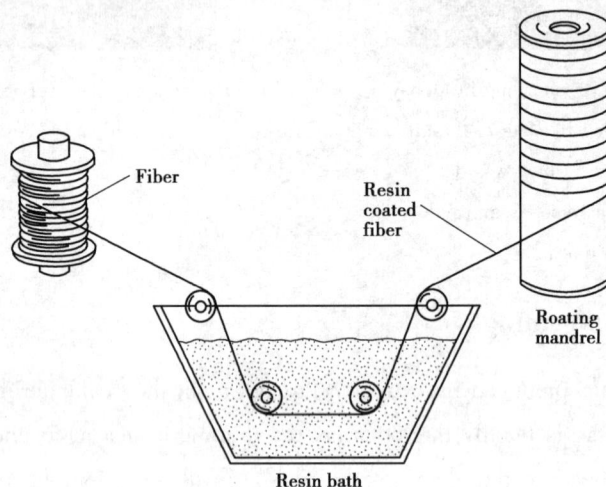

Figure 2.26　Illustration of the filament winding technique

2.3.12　Conclusion

A review of the many polymers and blends available and the many fabrication techniques that can be used to produce finished articles should demonstrate the versatility of polymers. Because of this versatility, the polymer industry has grown at an unprecedented rate and will be the world's largest and most important industry into the beginning of the twenty-first century.

2.4　Polyethylene[4-7]

Despite ethylene's simple structure, the field of polyethylene is a complex one with a very wide range of types and many different manufacturing processes. From a comparatively late start, polyethylene production has increased rapidly to make polyethylene the major tonnage plastics material worldwide (45×10^6 t capacity in 1995). In the 1920s research into the polymerization of unsaturated compounds such as vinyl chloride, vinyl acetate, and styrene led to industrial processes being introduced in the 1930s, but the use of the same techniques with ethylene did not lead to high polymers. The chance observation in 1933 by an ICI research team that traces of a waxy polymer were formed when ethylene and benzaldehyde were subjected to a temperature of 170 ℃ and a pressure of 190 MPa, led to the first patent in 1936 and small-scale production in 1939. The polymers made in this way, by using free radical initiators, were partially crystalline, and measurement of the density of the product was quickly established as a means of determining the crystallinity. Due to the side reactions occurring at the high temperatures employed, the polymer chains were branched, and densities of $915 \sim 925$ kg/m^3 were typically obtained. The densities of completely amorphous and completely crystalline polyethylene would be 880 and 1 000 kg/m^3, respectively.

During the 1950s three research groups working independently discovered three different catalysts which allowed the production of essentially linear polyethylene at low pressure and temperature. These polymers had densities in the region of 960 kg/m^3, and became known as high-density polyethylenes (HDPE), in contrast to the polymers produced by the extensively commercialized high-pressure process, which were named low-density polyethylenes (LDPE). These discoveries laid the basis for the coordination catalysis of ethylene polymerization, which has continued to diversify. Of the three discoveries at Standard Oil (Indiana), Phillips Petroleum, and by Karl Ziegler at the Max-Planck-Institute, the latter two have been extensively commercialized. More recently the observation that traces of water can dramatically increase the polymerization rate of certain Ziegler catalysts has led to major developments in soluble coordination catalysts and later their supported variants.

The coordination catalysts allowed for the first time the copolymerization of ethylene with other olefins such as butene, which by introducing side branches reduces the crystallinity and allows a low-density polyethylene to be produced at comparatively low pressures. Although DuPont of Canada introduced such a process in 1960, worldwide the products remained a small-volume specialty until 1978 when Union Carbide announced their Unipol process and coined the name linear low-density polyethylene (LLDPE). In addition to developing a cheaper production process, Union Carbide introduced the concept of exploiting the different molecular structure of the linear product to make tougher film. Following this lead, LLDPE processes have been introduced by many other manufacturers.

Figure 2.27 shows schematic structures for the three polyethylenes, with the main features exaggerated for emphasis. LDPE has a random long-branching structure, with branches on branches. The short branches are not uniform in length but are mainly four or two carbon atoms long. The ethyl branches probably occur in pairs, and there may be some clustering of other branches. The molecular mass distribution (MMD) is moderately broad.

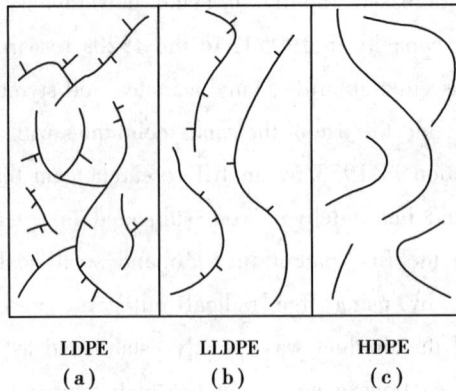

LDPE   LLDPE   HDPE

(a)   (b)   (c)

Figure 2.27   Schematic molecular structure

(a) Low-density polyethylene; (b) Linear low-density polyethlene;

(c) High-density polyethylene

LLDPE has branching of uniform length which is randomly distributed along a given chain, but there is a spread of average concentrations between chains, the highest concentrations of branches being generally in the shorter chains. The catalysts used to minimize this effect generally also produce fairly narrow MMDs.

HDPE is essentially free of both long and short branching, although very small amounts may be deliberately incorporated to achieve specific product targets. The MMD depends on the catalyst type but is typically of medium width.

Polyethylene crystallizes in the form of platelets (lamellae) with a unit cell similar to that of low molecular mass paraffin waxes. Due to chain folding, the molecular axes are oriented perpendicular to the longest dimension of the lamella and not parallel to it as might be expected (see Figure 2.28). The thickness of the lamellae is determined by the crystallization conditions and the concentration of branches and is typically in the range of 8 ~ 20 nm. Thicker lamellae are associated with higher melting points and higher overall crystallinities. Slow cooling from the melt or annealing just below the melting point produces thicker lamellae. Where long molecules emerge from the lamella they may either loop back elsewhere into the same lamella or crystallize in one or more adjacent lamellae, thereby forming "tie molecules".

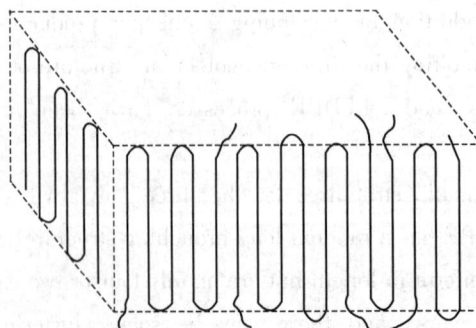

Figure 2.28   Folded-chain lamellae crystal of polyethylene

Thermodynamically the side branches are excluded from the crystalline region because their geometry is too different from that of the main chains to enter the crystalline lamellae. Therefore, the branches initiate chain folding, which results in thinner lamellae with the branches mainly situated on the chain folds on the surface of the lamellae. However, on rapid cooling these energetically preferred placements may not always occur, and some branches may become incorporated as crystal defects in the crystalline regions. Detailed measurements by solid-state NMR and Raman spectroscopy show that the categorization into crystalline and amorphous phases is too simplistic and a significant fraction of the polymer is present in the form of an "interfacial" fraction which has neither the freedom of motion of a liquid, nor the well-defined order of a crystal. A further result of a side branch is that having been prevented from folding directly into the same lamella, the polymer chain may form a tie molecule that links to one or more further lamellae.

Under moderately slow cooling conditions, crystallization may be nucleated at a comparatively small number of sites. Crystallization then propagates outwards from these centers until the surfaces of the growing spheres meet. The resulting spherulites show a characteristic banded structure under a polarizing optical microscope. The typical milkiness of polyethylene is due to light scattered by spherulites or other, less well defined aggregates of crystallites, rather than by the crystallites themselves, which are much smaller than the wavelength of light. Ethylene copolymers may be transparent, although partially crystalline.

LDPE and LLDPE are translucent whitish solids and are fairly flexible. In the form of films they have a limp feel and are transparent with only a slight milkiness. HDPE on the other hand is a white opaque solid that is more rigid and forms films which have a more turbid appearance and a crisp feel. The physical properties of LDPE, HDPE, LLDPE are compared in Table 2.4.

Table 2.4   Properties of some typical polyethylenes (data from Repsol Quimica)

Property	LDPE	HDPE	LLDPE	Method	Standard
Polymer grade	Repsol PE077/A	Hoechst GD-4755	BP LL 0209		
Melt flow index (MFI), g/600 s	1.1	1.1	0.85	190 ℃/ 2.16 kg	ASTM D1238
High load MFI, g/600 s	57.9	50.3	24.8	190 ℃/ 21.6 kg	ASTM D1238
Die swell ratio (SR)	1.43	1.46	1.11		
Density, kg/m³	924.3	961.0	922.0	slow annealed	ASTM D1505
Crystallinity, %	40	67	40	DSC	
Temperature of fusion (max.), ℃	110	131	122	DSC	
Vicat softening point, ℃	93	127	101	5 ℃/h	ASTM D1525
Short branches **	23	1.2	26	IR	ASTM D2238

continued

Property	LDPE	HDPE	LLDPE	Method	Standard
Comonomer		butene	butene	NMR	
Molecular mass *					
M_w	200 000	136 300	158 100	SEC	
M_n	44 200	18 400	35 800	SEC	
Tensile yield strength, MPa	12.4	26.5	10.3	50 mm/min	ASTM D638
Tensile rupture strength, MPa	12.0	21.1	25.3		
Elongation at rupture, %	653	906	811		
Modulus of elasticity, MPa	240	885	199	flexure	ASTM D790
Impact energy, unnotched, kJ/m^2	74	187	72		ASTM D256
notched, kJ/m^2	61	5	63		ASTM D256
Permittivity at 1 MHz	2.28				ASTM D1531
Loss tangent at 1 MHz	100×10^{-4}				ASTM D1531
Volume resistivity, W · m	10^{16}				
Dielectric strength, kV/mm	20				

* Corrected for effects of long branching by on-line viscometry

** Number of methyl groups per 1 000 carbon atoms

Polyethylene does not dissolve in any solvent at room temperature, but dissolves readily in aromatic and chlorinated hydrocarbons above its melting point. On cooling, the solutions tend to form gels which are difficult to filter. Although LDPE and LLDPE do not dissolve at room temperature, they may swell in certain solvents with deterioration in mechanical strength. Manufacturers issue data sheets detailing the suitability of their products for use in contact with a wide range of materials. In addition to solvents, polyethylene is also susceptible to surface active agents which encourage the formation of cracks in stressed areas over prolonged periods of exposure. This phenomenon, known as environmental stress cracking (ESC), is believed to be due to lowering of the crack propagation energy. In general, HDPE is the preferred polyethylene for liquid containers.

LDPE (together with LLDPE which is sold into the same market) is used predominantly for films, not all of which is for packaging. Because of its greater rigidity and better creep properties,

HDPE is used in more structural applications, and also has important applications in the packaging of aggressive liquids such as bleach, detergent, and hydrocarbons[8].

LDPE retains its position as a preferred packaging material because of its limp feel, transparency, toughness, and the ability to rapidly take up the shape of the contents of the bag. Other materials such as thinner HDPE film or paper may in some cases be more economical, but are less acceptable to the customer. Most LDPE film is produced by the film blowing process but flat film extruded onto chilled rolls is also made, particularly in the United States. An LDPE for high-clarity film typically has a MFI of 2 and a density of 920 kg/m^3 and is extruded at 160~180 ℃. The cast film process is usually used with higher density polyethylenes (930~935 kg/m^3), where the quenching on the chilled rolls enables good optical properties to be achieved with a higher film stiffness. The bubble diameter in the tubular film process may be up to 2 m for general purpose packaging, and larger for heavy gauge industrial film. Originally, bags were made directly from tubular film by welding one end, but the tendency now is to make wide film on large machines and fabricate the bags by welding and cutting. Apart from packaging film and heavy duty sacks, increasing quantities of polyethylene are used for impermeable or stabilizing membranes in civil engineering construction.

Additives play a particularly important role in the production of LDPE film by the film blowing process. Without additives, the pressure of the windup rollers on the warm film forces the surfaces into such close contact that subsequently it may be virtually impossible to separate them. High-gloss films are the worst affected. This overcome by adding an antiblocking agent such as very fine silica, which roughens the surface on a submicroscopic scale without significantly affecting the optical properties. To reduce the friction between the surfaces a slip agent such as oleamide or erucamide is added. Other additives may be added to achieve effects in the final product such as oxidation resistance, UV resistance, or antistatic properties.

A film extruder designed for LDPE requires extensive modifications to allow it to extrude LLDPE at comparable rates. To avoid this investment and the lack of flexibility, many film manufacturers (particularly in Europe) use blends of LDPE and LLDPE. The equipment modifications required are:

①The screw must have greater clearance between the flights and the barrel to avoid temperature buildup due to the higher shear viscosity. Ideally the screw should be shorter.

②The die gap must be widened to reduce the shear rate and so avoid a type of surface defect, known as shark skin, to which narrow MMD polymers are prone. To some extent this problem can be avoided by incorporating additives based on fluoroelas-towers (DuPont, 3M) or silicones (UCC).

③The cooling ring must give more rapid quenching of the melt to avoid the development of excessive crystalline haze, and also to give more support to the bubble because of the lower tensile viscosity under the inflation conditions. Apart from this lack of inherent stability of the bubble, the low tensile viscosity is responsible for the major product advantage of LLDPE, since high stresses do not occur when the melt is drawn down to thin film. LLDPE can be drawn down to much thinner film without the bubble tearing than can a LDPE with equivalent mechanical properties.

HDPE film is generally made on units specifically optimized for the high molecular mass grades normally used. To produce tough films a balanced orientation of the film is necessary, and a high ratio of die diameter to bubble diameter ("blow ratio") is used so as to balance the machine direction draw with a high transverse draw. A characteristic stalk-shaped bubble is used in which substantial machine-direction draw takes place before a rapid expansion occurs some distance above the die. Mechanical guides are needed to stabilize the bubble. HDPE film is opaque and is usually used in much thinner gauges than LDPE. It competes with LDPE in areas such as carrier bags and supermarket convenience bags.

HDPE is now the preferred material for blow-molded containers for liquids, combining adequate environmental stress crack resistance with higher rigidity than LDPE, and hence permitting lower bottle weights for a given duty. The major uses are in the domestic market for bleach, detergents, and milk. For this application a HDPE with a MFI of 0.2 and a density of 950 g/m^3 is suitable. Other uses include a variety of industrial containers and gasoline tanks. There is a trend to higher molecular mass and to broader MMD polymers, with bimodal MMDs achieved by using catalyst or reactor developments.

Injection molding is used for a variety of products. Some, such as caps and lids are used in the packaging field. Other applications such as housewares, toys, and industrial containers are more durable. According to whether high flexibility is required or not, LDPE or HDPE may be used. Some LLDPEs with a narrow MMD are particularly well suited for this method of fabrication. MFIs range from 2 to >40. A narrow MMD gives the best compromise between toughness, flow into the mold, and freedom from warpage in the finished product.

The use of polyethylene for pipes is one of the few engineering applications, where the applied stresses are carefully assessed and a lifetime of at least 50 years is required. Polyethylene is used for water and natural gas local distribution systems and, unlike many other uses, there is a substantial stress applied continuously. The potential problem is thus long-term stress rupture, and this must be carefully assessed by extrapolation and accelerated testing. Large quantities of pipe made of a medium-density polyethylene designed to withstand a stress in the PE of 8 MPa have been laid, but recent advances in bimodal HDPE polymers now enable a 50-year lifetime at a design stress of 10 MPa to be met. The first polymers which met this specification (PE 100) were introduced by Solvay in their TUB 120 range, but now all major producers have similar polymers in their ranges.

Because of its outstanding dielectric properties, the first application of polyethylene was the insulation of very high frequency and submarine telephone cables. These have continued to be made from LDPE, but uses have expanded into telephone and power cables. In the latter case cross-linking is increasingly used to enhance the properties and allow a higher current rating for a given size of cable.

2.5   Ethylene copolymers

In addition to copolymers of 1-olefins, such as LLDPE, there are several other commercial copolymers of ethylene. The copolymer of ethylene and vinyl acetate is an amorphous copolymer that may be cast as a clear film or used as a melt coating. The copolymer of ethylene and methacrylic acid ($CH_2 = C(CH_3)COOH$) is also a moldable thermoplastic. This copolymer, when partially neutralized to form monovalent and divalent metal-containing materials, is called an ionomer. These ionomer salts have a stable cross-linked structure at ordinary temperatures but can be injection molded. These tough copolymers are used as golf ball covers in place of balata.

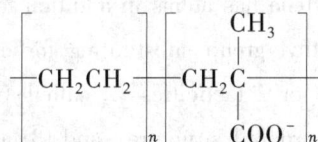

$$\left[CH_2CH_2 \right]_n \left[\begin{array}{c} CH_3 \\ | \\ CH_2C \\ | \\ COO^- \end{array} \right]_n$$

Ethylene-methacrylic acid copolymers (ionomers)

Both HDPE and polypropylene are high-melting crystalline polymers. However, the random copolymer of these two comonomers is an amorphous, low-melting elastomer. It is customary to add a cross-linking monomer, such as dicyclopentadiene, to the comonomers to produce a vulcanizable elastomer (EPDM). EPDM is used as the white sidewalls of tires and as single-ply roofing material.

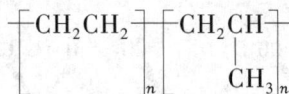

$$\left[CH_2CH_2 \right]_n \left[\begin{array}{c} CH_2CH \\ | \\ CH_3 \end{array} \right]_n$$

Ethylene-propylene copolymer

These ethylene-propylene copolymers are also employed in other automotive applications such as radiator and heater hoses, seals, mats, weather strips, bumpers, and body parts. Non-automotive applications include coated fabrics, gaskets and seals, hoses and wire, and cable insulators.

The block copolymer of ethylene and propylene, which contains long sequences of ethylene and propylene repeating units, is a clear, moldable copolymer and is used in place of HDPE in many applications. Its specific gravity is similar to that of LDPE.

2.6   Polypropylene

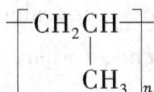

$$\left[\begin{array}{c} CH_2CH \\ | \\ CH_3 \end{array} \right]_n$$

Polypropylene (PP)

Nobel laureate K. Ziegler patented HDPE but failed to include polypropylene (PP) in his patent application. However, many other chemists used the Ziegler catalyst ($TiCl_3Al(C_2H_5)_3$) to produce PP in the early 1950s. Nobel laureate G. Natta of Montedison, W. Baxter of DuPont, and E. Vanderburg of Hercules filed for patents for the production of PP using the Ziegler catalyst. J. Hogan and R. Banks of Phillips and A. Zletz of Amoco filed for patents using supported metal oxide catalysts. In 1973 the U.S. Patent Office granted a patent for PP to Natta, but reversed its decision in favor of Hogan and Banks in 1983[9].

Polypropylene (PP) is one of the three most heavily produced synthetic polymers. This abundance of PP is called for because of its variety and versatility being employed today in such diverse applications as a film in disposable diapers and hospital gowns to geotextile liners; plastic applications as disposable food containers and automotive components; and fiber applications such as carpets, furniture fabrics, and twine.

Unlike polyethylene, polypropylene has atoms in addition to hydrogen attached to the polymer backbone. The presence of the methyl group substituting for one of the hydrogens gives rise to several different structural isomers or "tacticities", called ① isotactic, and ② syndiotactic isomers, which are both regular or ordered structure, and ③ atactic isomers, which have only a somewhat random arrangement of methyl groups (Figure 2.29).

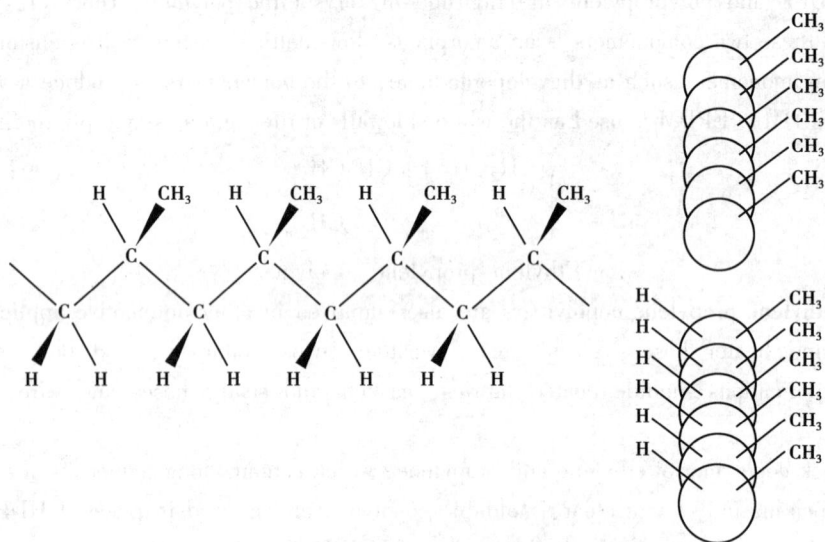

Figure 2.29   Illustrations representing isotactic polypropylene

The regular structures, isotactic and syndiotactic, allow the polymer chains to more readily come closer together, forming crystalline structures. These crystalline structures are reflected in so-called stereoregular PP being stronger, less permeable to acids, oils, and gas molecules, and higher-melting. The disordered random atactic, (where the prefix "a" means having nothing to do with) structure gives a less crystalline product.

Notwithstanding the complexities and hazardous nature of these new coordinated catalysts, industrialists and academe alike were stimulated by Ziegler and Natta's prompt and prolific disclosures concerning this new, high-melting polymer. Industrial-scale production of PP started at

Ferrara, Italy, in 1957 with a 6×10^3 t/a plant. By the end of 1994 global capacity had risen to 20.5×10^3 t/a (Table 2.5), spread over 35 countries, today, new plants are expected to be capable of producing 200×10^3 t/a to achieve reasonable economies of scale.

Table 2.5   Global PP capacity (10^3t/a) and its distribution

Country/region	End 1990	End 1994	Growth * 1990-1994/%
Western Europe	4 530	5 705	26
North America	4 400	5 440	24
Japan	1 870	2 656	42
Other Far East/Australia	2 734	4 090	50
Middle East/Africa	280	655	34
South America	450	1 035	30
Eastern Europe	790	930	18
Total/average	15 054	20 511	36

* 7% per annum growth for four years is 31% total growth.

There are several reasons for this phenomenal growth, which exceeds that of other bulk plastics (Table 2.6). At first propene was readily available, almost at byproduct prices, from petrochemical cracker plants making ethylene. The polymer itself was suitable for a wide range of existing and new applications such as films, various fibers, large and small moldings from boat hulls to instrument parts. In addition, new manufacturing technologies based on $MgCl_2$ supported catalyst systems yielded both cost savings and improved products by eliminating atactic PP removal steps and then the deashing stage. Some 40 years on there are expectations that the new metallocene catalyts will extend these uses, even to the extent of challenging products such as polyamide, ABS, and flexible PVC for part of their market.

Table 2.6   Major polymer usage in Western Europe

Polymer	Consumption/10^3t/a			Growth rate/ %	
	1974	1984	1994	'74/'84	'84/'94
LLDPE + LDPE	3 080	4 089	6 015	33	47
HDPE	1 035	1 704	3 570	65	110
PS	1 118	1 290	1 908	15	48
PVC	3 360	3 825	5 480	14	43
PP	680	1 895	5 178	179	173

Compounding includes incorporating mineral fillers, glass fibers, elastomer, flame retardants, pigments, and carbon black. It is a specialized operation dealing with relatively short run lengths in specified compounding extruders or mixers. For this reason, and for concern about cross contaminating unmodified PP grades, compounding plants usually operate separately from main PP production plants.

Substantial property enhancements enable polypropylene-based products to compete in quite demanding areas where unmodified PP would be inadequate. Examples include tough front and rear bumpers for cars, rigid coupled glass compositions for washing machine tubs, dense grades for outdoor garden furniture, and temperature-resistant compositions for car under hood applications. Long-fiber-reinforced PP secures further increases in toughness and stiffness over short-glass-fiber compounds. Fibers with a minimum length of between 0.8 and 3 mm, depending on the coupling between matrix and fiber, are likely to be satisfactory. These form an interlocking skeletal structure which dissipates the impact energy over a large area of the molding.

The combination of lower production rates, specialized equipment, and dense additives mean that PP compounds are always more expensive than natural grades.

Polypropylene is readily processed in conventional equipment used for other thermoplastics. Injection molding, commonly using screw-type reciprocating machines, accounts for $40\% \sim 50\%$ of all applications. Extrusion processes account for the remainder with domination by fiber and film.

An exceptionally wide range of injection-molded products stems from the rigidity, toughness, and chemical and temperature resistance of PP. Examples include automotive trim and ventilation components, bottle crates, industrial containers, washing machine tops and tubs, kitchenware, tool handles, domestic waste systems, and small boat hulls. Such a variety of articles requires careful grade selection according to the required impact strength of the product and the melt flow constraints of the processing equipment. An MFI ($230\ ℃/2.16\ \mathrm{kg}$) range of $2 \sim 20\ \mathrm{dg/min}$ meets most needs. Increasing the MFI not only assists mold filling and reduces cycle times, but it helps manufacture of complex moldings having high flow ratio geometries (flow ratio is the ratio of longest path to the section thickness). The inevitable fall in impact strength at high MFI can be recouped, in some cases, by moving from homopolymer to impact copolymer grades.

While molding conditions vary with the size and complexity of the article, the following machine settings will cover most products:

Melt temperature: $230 \sim 275\ ℃$

Injection pressure: $55 \sim 100\ \mathrm{MPa}$

Mold temperature: $40 \sim 80\ ℃$, lower still for thin walled items

The low moisture absorption of PP largely dispenses with any need to dry granular feedstock. Neither do gaseous or carbonaceous decomposition products form in extruder dead spots. Instead, polymer remaining there becomes progressively more fluid through chain scission. Machine shot weight capacities normally refer to performance with polystyrene, i.e., the polystyrene yardstick figure. The lower melt density of PP ($0.75\ \mathrm{g/cm^3}$) lowers the machine's weight capacity by ca. 25% when molding PP. Plasticizing capacity is slightly greater than for polyethylene, but is lower than for amorphous polystyrene. Machine outputs with PP benefit from the ability to eject moldings at higher temperatures than with their lower softening counterparts.

Linear post-molding shrinkage of PP is $1\% \sim 2\%$, of which 85% occurs within the first 24h. Distortion in molded products reflects internal stress caused by nonuniform cooling and polymer orientation. The latter is especially pronounced in surface skins where the viscous melt stretches as it is pulled along by the advancing melt front adjacent to cool walls. Controlled rheology (CR) polymer, containing fewer of the orientation-prone long chains, reduces this type of warping.

An important practical aspect of PP concerns its ability to form strong integral hinges, such as in a lidded box, where the lid is permanently attached to the base by a thin web of polymer along the whole of one edge (Figure 2.30). This web, about $0.25 \sim 0.6$ mm thick, can be produced directly in the molding process or by post forming operations. An initial flexing of the hinge while slightly warm induces the correct molecular orientation to permit repeated opening, even at subzero temperatures. In the laboratory, such hinges have withstood 23×10^6 flexes without failure.

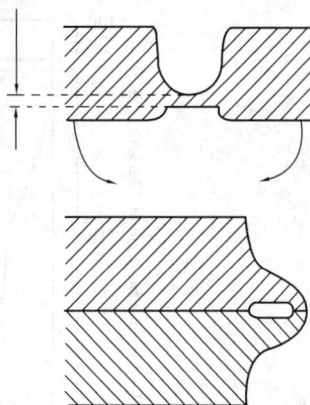

Figure 2.30   Integral hinge

Because polypropylene (PP) is low in cost but has outstanding mechanical properties and moldability, it accounts for more than half of all the plastic materials used in automobiles. PP compounds are used for a variety of parts, including bumper facias, instrumental panels and door trims. Several grades of PP compounds, with their diverse performance characteristics, have been developed by compounding PP with various other materials according to the performance requirements of the intended parts. As of 2007, 3.75 million tons of PP (8% of the world's total PP consumption of 45.5 million tons) and 690 000 tons of PP (approximately 23% of the total domestic PP consumption of 2.94 million tons) are used for automotive applications[10].

The growth of PP compounds for automotive applications has thus far been supported by the improved performance of PP resins—which serve as the base of PP compounds—and advancements in compound technology. With respect to the former, catalysts and the polymerization process have been continually, energetically improved in order to control the primary and higher-order structures of polymers. Regarding the latter, improvements in the performance and dispersibility of elastomers, as well as the control of particle size, dispersion and interface of inorganic fillers, have been attempted up to the present time.

With respect to resin-based automotive parts, lower weight is demanded for the sake of reduced environmental burden and better design and higher moldability are also required. In response to that demand, various phases of PP compounds for automotive applications have been improved. The improvements made thus far include greater rigidity, impact strength, fluidity and crystallization. Such enhancements of PP compounds have been achieved by compounding PP with additives such as elastomers and/or various inorganic fillers, as well as through higher stereoregularity, fluidity and rubberization which have been achieved with the aforementioned improvements in catalysts and the manufacturing process. A wide range of performance requirements in automotive parts, for which various engineering plastics were conventionally used, can now be covered by PP compounds alone (Figure 2.31). As a result, PP-based material consumption in automotive applications has continued to increase.

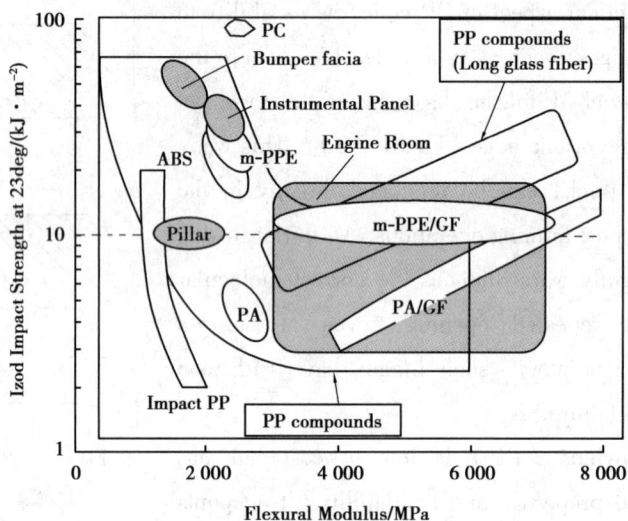

Figure 2.31　Mechanical properties of PP compounds

To improve the rigidity of PP, material development has been conducted using short-length glass fiber. By adding short-length glass fiber to PP, the threshold of heat resistance increases to a point near the melting point of PP. This type of PP is therefore used in automotive parts for areas to be exposed to severe heat, such as the engine compartment. Furthermore, in order to add even more rigidity, long glass fiber reinforced PP is currently under development. As shown in Figure 2.31, the long glass fiber reinforced PP demonstrates an extremely high flexural modulus. Through this system a material having extremely high tensile strength can be obtained by extending the residual fiber length.

The short glass fiber reinforced PP is produced using a standard type of twin-screw extruder. However, in order to prevent the glass fiber in the compound from breaking, the side-feed method is used, by which glass fiber is added to the molten resin from the side of the extruder. On the contrary, the long glass fiber reinforced PP is produced according to the following procedures: Glass fiber is continuously supplied from glass-fiber rolls, in a manner called roving, to the molten resin in the impregnating die. The glass fiber is then coated with the molten resin. Lastly, the fiber is cut into 5~40 mm length (Figure 2.32). Thus the pellet length and glass fiber length are equal in the long glass fiber reinforced PP. The long-glass-fiber reinforced PP is used for front-end modules, back door panels and door interior panels for automotive parts. In order to satisfy the performance requirements, it is important to have the technology that enhances the strength of interface between the glass fiber and PP. Furthermore, because it is important to improve the dispersion of glass fiber for exterior parts, materials having excellent glass fiber dispersion suitable for this purpose are now being developed through an improved manufacturing method and the optimization of glass fiber. Regarding other fiber reinforced PP, carbon fiber reinforced PP and environmentally friendly organic fiber reinforced PP are currently under development.

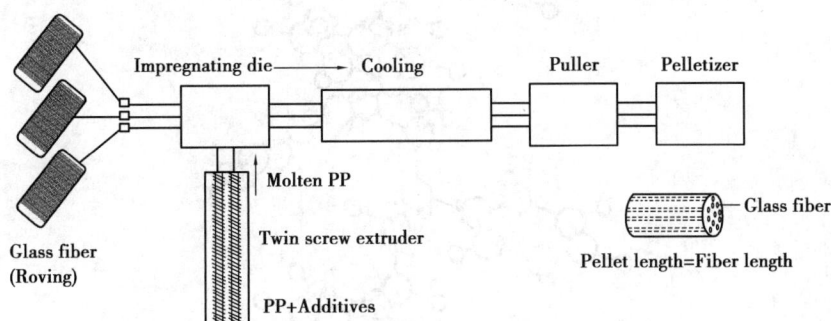

Figure 2.32   Process for long glass fiber reinforced PP

2.7   Polystyrene

Styrene monomer was discovered by Newman in 1786. The initial formation of polystyrene was by Simon in 1839. While polystyrene was formed almost 175 years ago, the mechanism of formation was not discovered until the early twentieth century. Staudinger, using styrene as the principle model, identified the general free radical polymerization process in 1920. Initially, commercialization of polystyrene, as in many cases, awaited the ready availability of the monomer. While there was available ethyl benzene, it underwent thermal cracking rather than dehydrogenation until the appropriate conditions and catalysts were discovered. Dow first successfully commercialized polystyrene formation in 1938. While most commercial polystyrene (PS) has only a low degree of stereoregularity, it is rigid and brittle because of the resistance of easy movement of the more bulky phenyl-containing units in comparison, for example, to the methyl-containing units of polypropylene (Figure 2.33). This is reflected in a relatively high T_g of about 100 ℃ for polystyrene. It is transparent because of the low degree of crystalline formation.

While PS is largely commercially produced using free radical polymerization, it can be produced by all four of the major techniques: anionic, cationic, free radical, and coordination-type systems. All of the tactic forms can be formed employing these systems. The most important of the tactic forms is syndiotactic PS (sPS). Metallocene-produced sPS is a semicrystalline material with a T_m of 270 ℃. It was initially produced by Dow in 1997 under the trade name of Questra. It has good

53

chemical and solvent resistance in contrast with "regular" PS, which has generally poor chemical and solvent resistance because of the presence of voids, due to the presence of the bulky phenyl groups, which are exploited by the solvents and chemicals.

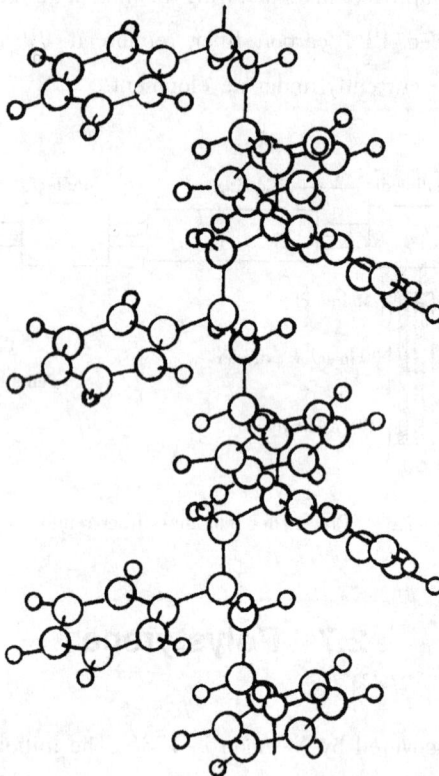

Figure 2.33 Ball-and-stick model of polystyrene

Physical properties of PS are dependent on the molecular weight and presence of additives. While higher-molecular-weight PS offers better strength and toughness, it also offers poorer processability. Low-molecular-weight PS allows for good processability but poorer strength and toughness. Generally a balance is sought where intermediate chain lengths are used.

Because PS is brittle with little impact resistance under normal operating conditions, early work was done to impart impact resistance. The best-known material from this work is called high-impact polystyrene or HIPS. HIPS is produced by dispersing small particles of butadiene rubber in with the styrene monomer. Polymerization captures the butadiene rubber particles within the polymerizing styrene.

Major uses of PS are in packaging and containers, toys and recreational equipment, insulation, disposable food containers, electrical items and electronics, housewares, and appliance parts. Expandable PS is used to package electronic equipment such as TV's, computers, and stereos.

Legislation was put in place in some states to ensure the recycling of PS. Interestingly some of this legislation was written such that all PS had to be recycled within some period of time such as a year. This legislation was changed to reflect the real concern of fast food containers when it was pointed out that less than 10% PS is used in this manner and that well over twice as much was used as house insulation that should not be recycled every year or so.

2.8   Styrene copolymers

In addition to the SBR elastomer described in Chapter 6, a less rubbery copolymer with a lower percentage of butadiene is used as a tough plastic. Styrene-acrylonitrile copolymers (SANs) have relatively high heat deflection temperatures. Because of their thermal stability, SANs are employed in the production of "dishwasher-safe" houseware, such as blender bowls, humidifier parts, detergent dispensers, and refrigerator vegetable and meat drawers. Also, fiberglass-reinforced automotive battery cases and dashboard components are molded from SAN.

Blends of SAN and butadiene-acrylonitrile rubber (NBR) have superior impact resistance. Sheets of this acrylonitrile-butadiene-styrene (ABS) terpolymer are thermoformed for the production of suitcases, crates, and appliance housings. The weather resistance and clarity of ABS is improved by replacing the acrylonitrile by methyl methacrylate.

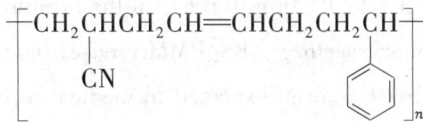

$$\left[CH_2CHCH_2CH{=}CHCH_2CH_2CH \right]_n$$

Acrylonitrile-butadiene-styrene terpolymer (ABS)

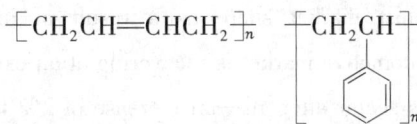

$$\left[CH_2CH{=}CHCH_2 \right]_n \quad \left[CH_2CH \right]_n$$

Styrene-butadiene rubber (SBR)

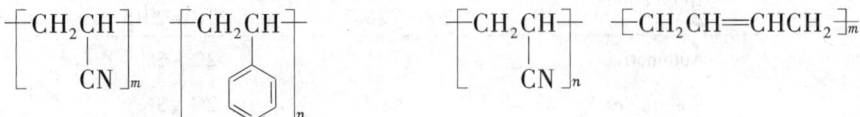

$$\left[CH_2CH \right]_m \quad \left[CH_2CH \right]_n \quad \left[CH_2CH \right]_n \quad \left[CH_2CH{=}CHCH_2 \right]_m$$

Styrene-acrylonitrile copolymer (SAN)   Nitrile-rubber (NBR)

Acrylonitrile-butadiene-styrene is still the most important of the styrene copolymers. It clearly dominates the entire market, followed by SAN, ABS blends, ASA and MABS The list of the ten ABS suppliers with the largest capacities (Table 2.7) is dominated by Asian manufacturers. ABS from Far-East sources now covers significantly more than half the world market[11].

Table 2.7 **The ten leading ABS producers in the world** (date: mid 2011; the styrene copolymers of BASF SE are now marketed under the name Styrolution)

Rank	Company	Capacity [million t]	Share/%
1	Chi Mei	1.490	17.20
2	LG Group	1.100	12.70
3	Formosa Group	0.710	8.19
4	BASF SE	0.677	7.81
5	Ineos	0.515	5.94
6	Samsung Group	0.430	4.96
7	Toray	0.422	4.87
8	Sabic	0.377	4.35
9	Grand Pacific	0.310	3.58
10	CNPC	0.306	3.53
Total		6.337	73.13

In the toy sector, for example, growth of 4% to 5% is expected in the coming five years (Table 2.8). The specialists believe extruded sheets and profiles (for the construction sector, among others) to be capable of 3% to 4% growth p.a.; in the furniture sector, there is a trend away from PVC and PP; in the sanitary sector, ABS/PMMA co-extrudates are increasingly replacing metals. Some 6% to 8% p.a. growth is even expected in medical technology. The material is used, for example, in housing parts, and is profiting from a strong trend towards cost-reducing self-medication. One of the most important consumers in recent years continued to be the automotive industry. Here, the material lost market share as a result of inter-polymer competition with polypropylene. However, the automotive market is recovering at an extremely promising rate, so that ABS is supposed to grow in this sector after all—an increase of 2% to 3% p.a. is expected.

Table 2.8 **Growth rates: ABS is still in demand in many markets**

Applications	Growth rates/%	
	2003~2007	2010~2015
Automotive	1%	2%~3%
Electronics	1%	2%~3%
Large appliances	−3%	1%~2%
Small appliances	2%	2%~3%
Sheet and profile extrusion	6%	3%~4%
Toys	5%	4%~5%
Medical	6%	6%~8%

BASF, Ludwigshafen, Germany, met this challenge by investing in a cost-efficient large plant at its Antwerp site in Belgium, where it restricted itself to producing unpigmented standard resin with an optimum price/performance ratio. The Kettler company is cladding its new cross trainer Satura P 7653 with Terluran® GP-35, a standard ABS (acrylonitrile-butadiene-styrene polymer) from BASF's product range of plastics, which Kettler itself then colors in light grey and black (Figure 2.34). This Terluran type is manufactured at the ABS plant in Antwerp, Belgium, which will be celebrating the production of its one millionth ton in July of 2009.

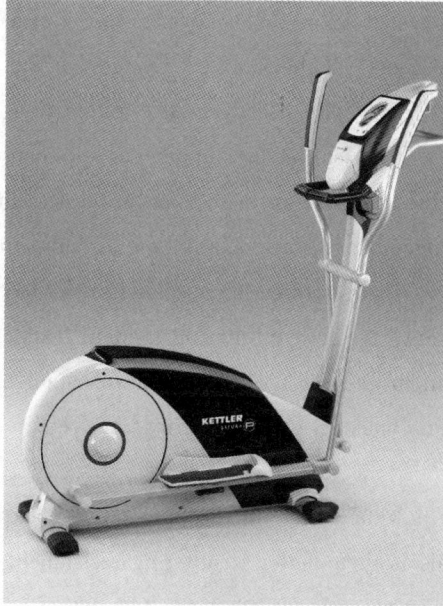

Figure 2.34   A commodity material, but still good to look at: the cross
trainer Satura P 7653 from Kettler is clad with a standard ABS;
Kettler pigments the resin itself and thereby saves time
and costs (photo: Kettler)

For self-coloring of ABS, BASF offers the system Colorflexx®, which was developed in cooperation with leading masterbatch suppliers in Europe in mid-2003. BASF supplies the undyed standard ABS, a so-called natural-colored product, while the masterbatch manufacturers provide the desired color concentrate. Working together, the suppliers ensure a smooth coloring operation at the customer's premises and, right on site, they solve any problems that might arise. The Colorflexx® model, which has already proven its worth for six years in Europe, goes hand in hand with the large world-scale ABS plant in Antwerp, Belgium, where only a few large-volume standard types of Terluran® are manufactured.

The advantages of this concept include inexpensive materials, efficient logistics, short delivery times, high availability, small warehouse inventories, low capital tie-up and resultant high flexibility, all aspects that are greatly appreciated by plastic processors, who are proceeding with caution as they look for economic solutions during these times of severe cost pressure.

New material developments in pure styrene copolymers are of course becoming ever more difficult. The situation is different for ABS blends, whose total market (together with corresponding ASA blends) is estimated at well over 500 kt/a. The possibilities resulting from the intelligent combination of different material components are by no means exhausted. Blends are therefore strategically capturing new market segments; thus their growth figures are regarded as exceptionally good.

Blends of ABS and polycarbonate, for example, still have considerable market importance. Large suppliers are Bayer Material Science AG, Leverkusen, Germany, (Bayblend) and Sabic Innovative Plastics, Bergen op Zoom, Netherlands, (Cycoloy). The main field of application for the amorphous thermoplastic (ABS+PC) blends is the automotive sector; experts estimate the volume of the European (ABS+ PC) market for automotive applications alone at about 50 000 to 60 000 t/a. In the exterior, the material is used, for example, in robust radiator grilles and (electroplatable) decorative elements, in the cockpit, for example in door trim, display consoles and safety parts, such as airbag covers. Here, their high heat deflection temperatures and impact values are particularly in demand; the low splintering tendency and good low-temperature impact resistance are particularly useful in case of a crash. Flame-resistant modified (ABS+ PC) blends are used in electrical equipment housings; the electrical and electronic sector is even the main growth segment.

(ASA+PC) blends are still widely used in the automotive cockpit, too. Their key advantage compared to (ABS+PC) blends is their better weathering resistance. Glass fiber (GF) reinforced grades have excellent flow properties even without large amounts of additives. The applications are, e. g., mirror covers and rearview mirror fairings. Sabic recently expanded its (ASA+PC) portfolio (Geloy) with new grades for use in medicine, entertainment electronics and electrical applications. For light but matt surfaces in automotive interiors, dust-repellent antistatic grades are available.

Blends of ABS and polyamides are characterized by an independent property profile comprising high impact resistance combined with good chemical resistance and good flow properties. The latter leads to good tool reproduction, allowing matt surfaces to be used without painting. In addition, they offer a range of acoustic and tactile properties; even glass fiber-reinforced parts, which already offer considerable stiffness with low GF admixtures, can be used for visible parts without painting (e.g., Terblend N NG-02). (ABS+PA) blends are therefore predominantly used in the automotive sector.

With Terblend N NM 21-EF, Styrolution has recently added an (ABS+PA) blend, which is characterized by even better flow properties. This pays off for the manufacture of large-area parts with a challenging wall thickness/flow path ratio. The new material thus meets the demands of the market for such applications and has found an interesting mass application in (unpainted) loud-speaker covers for a major Ger man automotive OEM (Figure 2.35). The intricate hole pattern in these covers would be difficult to realize with other materials.

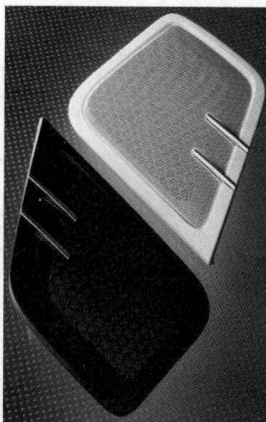

Figure 2.35 An enhanced-flow (ABS+PA) blend (Terblend N
NM 21-EF) has found a first mass application in automotive
interiors as loudspeaker grills (photo: Styrolution)[12]

2.9 Polyvinyl chloride and copolymers[13]

Polyvinyl chloride (also called vinyl or simply PVC) is a versatile thermoplastic material that is used in the production of hundreds of products that consumers encounter in everyday life and many more that are encountered less frequently but are nevertheless very important in construction, electronics, healthcare, and other applications. It finds widespread use in these applications because of its low cost and desirable physical and mechanical properties. It is fabricated efficiently into a very wide range of both rigid and flexible products. PVC also has inherent flame resistance. Substitutes for PVC materials may be available, but often the alternative materials and processes are not as efficient or substitution costs are high. It is the most widely consumed thermoplastic material after the polyolefins (polyethylene and polypropylene). Consumption in 2007 in the United States and Canada amounted to over 6.4 million metric tons, or almost 14.2 billion pounds (see Figure 2.36).

Consumers select products that contain PVC even when lower cost substitutes are available because they offer longer lives and reduced maintenance costs or aesthetic appeal. A breakdown of consumption of PVC resin (the powdered form of pure PVC) by major end use application and a compilation of some of the many products made from PVC is presented in Figure 2.36.

PVC has found widespread use because of its desirable properties, low cost and versatility as shown in Table 2.9. It starts as a powder that is derived from salt and fossil fuel. The ability to manipulate its characteristics through selection of the appropriate manufacturing and fabrication processes, and use of appropriate additives, including plasticizers for flexible products, is unmatched by any other thermoplastic material.

PVC resin consumption in 2007 totaled 6 426 thousand metric tons

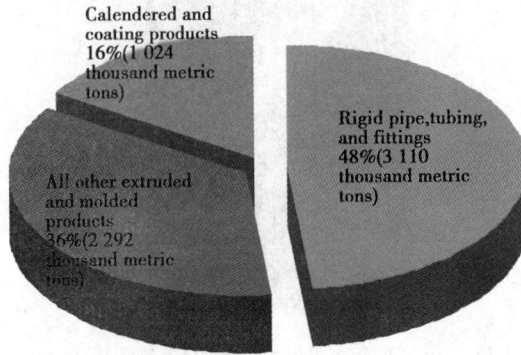

Calendered and
coating products
16%(1 024
thousand metric
tons)

Rigid pipe,tubing,
and fittings
48%(3 110
thousand metric
tons)

All other extruded
and molded
products
36%(2 292
thousand metric
tons)

Figure 2.36   Consumption of PVC Resin in the United

States and Canada, 2007

Table 2.9   Some representative products manufactured from PVC

CONSTRUCTION	AUTOMOTIVE
Piping & fittings for water distribution, irrigation & sewers; grey water recycling kits; electrical conduits; siding, awnings, soffit, skirting, weather stripping, gutters & downspouts; decking & fencing; window, door frames and cladding; landfill liners & geomembranes; swimming pool liners; single-ply roofing; conveyor belts; piping used in food processing, chemical processing & other manufacturing; floor & wall coverings; coated paneling; adhesives; maintenance coatings	Interior upholstery; "soft" dashboard & arm rests; dashboard instrument components, airbag covers; body side moldings, bumper guards; windshield system components, rearview mirror housings; under-the-hood wiring; under-the-car abrasion coatings; floor mats; adhesives & sealants; boots & bellows; battery separators; audio & video components; lighting components; steering cover & transmission parts; A/C system components
MEDICAL & HEALTHCARE	**ELECTRICAL AND ELECTRONICS**
Blood bags and tubing; cannulae; caps; catheters; connectors; cushioning products; device packages; dialysis equipment & tubing; drainage tubing; drip chambers; ear protection; goggles; inflatable splints; inhalation masks; IV containers and components; laboratory ware; masks; mouthpieces; oxygen delivery components; seals; surgical wire; jacketing; thermal blankets; urine and colostomy bags; valves and fittings	Computer housing & cabling; printed circuit board trays; power wire insulation & sheathing; communication cable jacketing; backing for power cable; electrical plugs & connectors, wall plates, connection boxes; soft keyboards; keyboard trays; coating for optical mouse pads; memory stick & USB covers/casings; LED product components; laminate for plastic security passes & "smart cards"
PACKAGING	**CONSUMER PRODUCTS AND OTHER**
Sterile medical packaging; tamper-proofing over-the-counter medication; shrink wrap for software, games, & household products; blister and clamshell packaging to protect toys, hardware, electronics, personal care products, & foods such as eggs and meat; bottles for household & personal care products, cooking oils & automotive lubricants; closures for bottles & jars; can coatings	Wind turbine blades; machinery parts; housings & handles for tools; garden hoses; tarpaulins; patio furniture, upholstery; appliance housings; window shades & blinds; table cloths, place mats, shower curtains; sporting goods, beach balls; vinyl leather goods; luggage, footwear, gloves, rainwear; handbags; apparel; coated paper; holiday decorations; toys

Source: Whitfield and Associates

Some products contain copolymers that incorporate other materials, such as vinyl acetate and vinylidene chloride, in order to impart desirable mechanical or processing properties. About 90% of PVC is produced by suspension polymerization in which the monomer is suspended in an aqueous mixture containing buffers, initiators, and colloid-forming agents, and reaction conditions are controlled to produce material with the desired molecular weight distribution, particle size, and particle morphology. The finished PVC resin is separated from the aqueous mixture, dried, and sold in powder form. PVC is made by other polymerization techniques when material with specific properties is required. For example, emulsion polymerization is used to produce plastisols, which are fluid dispersions of PVC in plasticizers, used to make calendered and coated products such as shower curtains and raincoats. Nonaqueous solution polymerization in organic solvents is used to produce specialty polymers and copolymers that are used in other coating and calendering applications such as floor tiles. Bulk polymerization is used to produce resins with high clarity for the manufacture of such products as blow molded bottles.

The PVC-containing products that consumers use every day are always produced from mixtures of PVC resin that has been compounded with other materials, often into pellet form, before being fabricated into their final forms. Rigid PVC products can contain from 10% to 20% by weight of various additives and fillers. The additives include stabilizers, pigments, impact modifiers, and processing aids that are used to facilitate the fabrication processes. PVC products can also incorporate significant amounts of inert filler materials that reduce product costs without compromising flexibility, toughness, and mechanical strength in many applications. Flexible PVC products may also contain additives for heat stabilization, UV absorbance, flame retardance, and lubrication. These products always contain significant amounts of plasticizers so that the fabricated article may contain as little as 50% PVC resin. Historically, phthalate plasticizers, particularly diethylhexyl phthalate (DEHP), have accounted for the great majority of plasticizer consumption in flexible PVC products. More recently, plasticizers based on adipates and other esters have been used in applications where they provide advantages in processing or performance.

PVC-containing products are manufactured from compounded resin by a number of processes, including extrusion, injection molding, calendering, coating, thermoforming, blow molding, and blown film processing. The most important of these are described below:

①Extrusion: a manufacturing process that uses a device similar to a pasta machine. The PVC resin is fed into the process and a screw-like shaft is rotated to push the resin forward where it is extruded through the outlet die. This technology is used to make pipe, conduit, siding, window frames, interior moldings, film, and sheet.

②Injection molding: a manufacturing process that injects PVC resin into a metal mold by pressure. This process is best suited for production of three-dimensional structures. This technology is used to produce pipe fittings, containers, and buckets.

③Calendering: a manufacturing process in which resin is heated and kneaded while passing through several pairs of rollers to be pressed to the required thickness. This technology is used to produce wide, flat products like floor tiles, geomembranes, artificial leather, and wall coverings.

④Coating: a manufacturing process that applies a PVC-based solution to one surface of PVC film or fabric and then dries it by heat to produce the final product. This technology is used to make large tents, connecting sections of trains, sign boards, and table cloths.

⑤Thermoforming: a manufacturing process in which PVC sheet or film (either extruded or calendered), is heated to soften it and pulled into a metal mold by a vacuum. This technology is used to make egg cartons, food trays, blister packaging, and press-through packaging for pills.

Both rigid and flexible forms of PVC can be extruded, but the equipment and processing techniques required are tailored to the properties of the compounded materials, and differ from those used with other thermoplastics. Injection and blow molding techniques are used to produce a variety of rigid products, while other molding techniques are used to produce flexible products. Calendering is used to produce both rigid and flexible products in film or sheet form, often with a backing material to form laminated products. PVC resins compounded into plastisols are applied to various substrates to produce a range of coated products for industrial and consumer applications.

In the following sections, we describe the use of PVC in the production of rigid pipe and fittings, all other extruded and molded products, and in calendered and coated products. We identify and describe the materials that might be substituted for PVC in the various applications and the issues involved in such substitution. Following this step, we estimate the direct costs of substituting alternate materials for PVC-containing products. These costs are the monetary benefits that consumers currently enjoy from access to PVC. Finally, we estimate the indirect or derivative benefits that PVC production brings to the economy.

As a side note, there is today a debate concerning the use of chlorine-containing materials and their effect on the atmosphere. This is a real concern and one that is being addressed by industry. PVC and other chloride-containing materials have in the past been simply disposed of through combustion that often created unwanted hydrogen chloride. This practice has largely been stopped, but care should be continued to see that such materials are disposed of properly. Furthermore, simply outlawing of all chloride-containing materials is not possible or practical. For instance, we need common table salt for life, and common table salt is sodium chloride. Chlorine is widely used as a water disinfectant both commercially (for our drinking water) and for pools. Furthermore, PVC is an important material that is not easily replaced. Finally, the amounts of chloride-containing residue that is introduced into the atmosphere naturally is large in comparison to that introduced by PVC. Even so, we must exercise care because we want to leave a better world for our children and grandchildren, so a knowledge-based approach must be taken.

Because of its versatility, some unique performance characteristics, ready availability, and low cost, poly(vinyl chloride), PVC, is now the second largest produced synthetic polymer behind polyethylene. PVC materials are often defined to contain 50% or more by weight vinyl chloride units. PVC is generally a mixture of a number of additives and often other units such as ethylene, propylene, vinylidene chloride, and vinyl acetate. In comparison to many other polymers, PVC employs an especially wide variety of additives. For instance, a sample recipe or formulation for common stiff PVC pipe such as used in housing and irrigation applications may contain (in addition to the PVC resin) tin stabilizer, acrylic processing aid, acrylic lubricant-processing aid, acrylic impact modifier, calcium carbonate, titanium dioxide, calcium stearate, and paraffin wax. Such formulations vary according to the intended processing and end use. In such nonflexible PVC materials the weight amount of additive is on the order of 5% ~ 10%.

PVC has a built-in advantage over many other polymers in that it is itself flame-resistant. About 50% of PVC is used as rigid pipe. Other uses of rigid PVC are as pipe fittings, electrical outlet boxes, and automotive parts. Uses of flexible PVC include in gasoline-resistant hose, hospital sheeting, shoe soles, electrical tape, stretch film, pool liners, vinyl-coated fabrics, roof coatings, refrigerator gaskets, floor sheeting, and electrical insulation and jacketing. A wide number of vinyl chloride copolymers are commercially used. Many vinyl floor tiles are copolymers of PVC.

Plastisol PVC products are made by heating to 150 ℃ finely divided PVC suspended in a liquid plasticizer that is in a mold.

In addition to its copolymer with vinyl acetate (Vinylite), vinyl chloride is also copolymerized with vinylidene chloride ($H_2C = CCl_2$) (Saran, Pliovic).

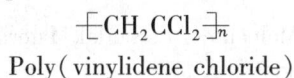

$$\text{---}CH_2CCl_2\text{---}_n$$
Poly(vinylidene chloride)

Collection of Exercises

1. What is metallocene PE? And what are the differences among metallocene PE, HDPE and LLDPE?
2. How many types of conformations do the polypropylene have?
3. What are the main properties of PS? When compared with PS, what advantage does ABS have?
4. Define Polyvinyls, PS, PP, HDPE, chemical structure.
5. Compare the density PVC, PVB, PS, and PVDC which is higher/lower than PP.
6. Compare the density of HDPE, LDPE, UHMWPE, LLDPE to PP?
7. What is the tensile strength of PP with 0%, 30% glass fibers? What is the tensile modulus?
8. Plot tensile strength and tensile modulus of PVC, PS, PP, LDPE and HPDE to look like:

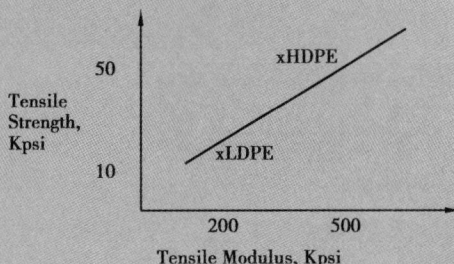

9. Which of the following are thermoplastics: hard rubber, Bakelite, PVC, polystyrene, polyethylene?

10. What is the principal structural difference between LDPE and HDPE?

11. Which has the higher value for a specific polymer with both amorphous and crystalline regions: T_g or T_m?

REFERENCES

[1] Rolf Klein, Laser Welding of Plastics, First Edition. Wiley-VCH Verlag GmbH & Co. KGaA, 2011.

[2] Charles E. Carraher Jr., Giant Molecules. Essential Materials for Everyday Living and Problem Solving, Wiley-Interscience, 2003.

[3] R.J. Crawford, Plastics Engineering, 3rd edition, Butterworth-Heinemann, 1998.

[4] Kenneth S. Whiteley, Ullmann's Encyclopedia of Industrial Chemistry (Polyethylene), Wiley, 2011.

[5] Charles A. Harper, Modern plastics handbook, McGraw-Hill Professional, 2000.

[6] http://www.rtpcompany.com/cn/index.htm, RTP 公司产品技术资料.

[7] http://www.lyondellbasell.com/Products/ByCategory/, Basell 公司产品技术资料.

[8] http://www.sasol.com/, Sasol 公司产品技术资料.

[9] Markus Gahleitner, Christian Paulik, Ullmann's Encyclopedia of Industrial Chemistry (Polypropylene), Wiley, 2011.

[10] Satoru Moritomi, Tsuyoshi Watanabe, Susumu Kanzaki, Polypropylene Compounds for Automotive Applications, Sumitomo Chemical Co., Ltd., 2010.

[11] Sabine Oepen, Axel Gottschalk, Styrene Copolymers (ABS, ASA, SAN, MABS, and ABS Blends), Carl Hanser Verlag, Munich, Kunststoffe International, (10): 22-26, 2011.

[12] http://worldaccount.basf.com/wa/plasticsEU ~ en_GB/portal, ABS/PA Terblend N with improved flow in speaker grilles, BASF 公司.

[13] The Economic Benefits of Polyvinyl Chloride in the United States and Canada, Whitfield and Associates based on American Chemistry Council, 2008.

CHAPTER 3

ENGINEERING PLASTICS

3.1 Introduction

Engineering plastics are those special polymers with excellent properties, which can be used as constructional materials, even in trenchant chemical and physical circumstance. In other words, it generally refers to those plastics which are able to bear certain external force, with good mechanical properties and dimensional stability, be able to maintain the excellent properties under both high temperature and low temperature and can be used as engineering components. It has greater Environmental bearing capacity and adaptability than universal plastics.

Engineering plastic can be divided into two categories: specially designed engineering plastic and general engineering plastic. General engineering plastic usually refers to five thermoplastic including polyamide (PA), polyformaldehyde (POM), polycarbonate (PC), modified polyphenylene oxide (PPO) and polyester (PET and PBT). Specially designed engineering plastic refers to engineering plastic that have greater properties other than the five above. When it comes to the operating temperature, usually, the operating temperature of general engineering plastic is below 150 ℃ (generally 100～150 ℃), the operating temperature of specially designed engineering plastic is above 150 ℃. The latter can be classified into 150～250 ℃ type (including compound of general engineering plastic) and above 250 ℃ type. The higher the operating temperature is, the higher the price would be. When it comes to application, engineering plastic can be divided into two parts. One is functional unit which has the requirement of strength, resistance to chemical reagents, abrasion resistance, etc. Another is mechanical part which has the requirement of shrinkage ratio, dimensional accuracy, appearance, etc. Generally speaking, the former is crystalline engineering plastic, the latter is noncrystalline engineering plastic. In the practical application, it often requires both of the properties. Applying class fiber (GF) and carbon fiber (CF) can be better adapted to this kind of requirement.

3.2 Nylons

Wallace Hume Carothers was brought to DuPont because his fellow researchers at Harvard and the University of Illinois called him the best synthetic chemist they knew. He started a program aimed at understanding the composition of natural polymers such as silk, cellulose, and rubber. Many of his efforts related to condensation polymers were based on his belief that if a monofunctional reactant reacted in a certain manner forming a small molecule, then similar reactions except employing reactants with two reactive groups would form polymers[1].

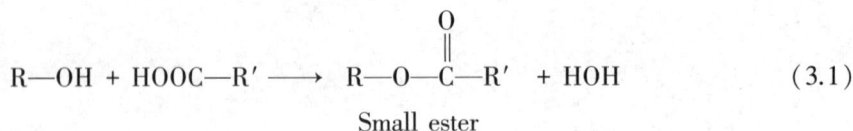

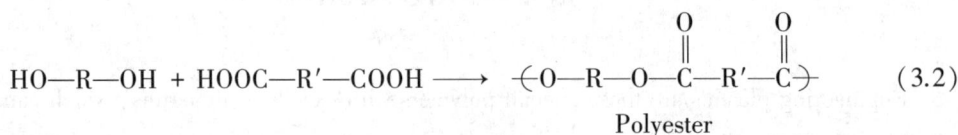

$$R-OH + HOOC-R' \longrightarrow R-O-\overset{\overset{\displaystyle O}{\|}}{C}-R' + HOH \qquad (3.1)$$

Small ester

$$HO-R-OH + HOOC-R'-COOH \longrightarrow \left(\!O-R-O-\overset{\overset{\displaystyle O}{\|}}{C}-R'-\overset{\overset{\displaystyle O}{\|}}{C}\!\right) \qquad (3.2)$$

Polyester

While the Carothers group had made both polyesters and polyamides, they initially emphasized work on the polyesters since they were more soluble and easier to work with. One of Carothers co-workers, Julian Hill, noticed that he could form fibers if he took a soft polyester material on a glass stirring rod and pulled some of it away from the clump. Because the polyesters had too low softening points for use as textiles, the group returned to work with the polyamides. They found that fibers could also be formed by the polyamides similar to those formed by the polyesters. These fibers allowed the formation of fibers that approached, and in some cases surpassed, the strength of natural fibers. This new miracle fiber was introduced at the 1939 New York World's Fair in an exhibit that announced the synthesis of this wonder fiber from "coal, air, and water"-an exaggeration but nevertheless eye-catching. These polyamides were given the name "nylons". When the polyamides, nylons, were first offered for sale in New York City on May 15, 1940, over 4 million pairs were sold in the first few hours. Nylon sales took a large drop when nylon was used to produce the parachute material so critical to WW II.

Although nylon 6,6 (Figure 3.1) and nylon 6 are used primarily as fibers, they are also used as engineering plastics. In fact, nylon 6,6 was the first engineering thermoplastic and until 1953 it represented the entire annual engineering thermoplastic sales. Nylon 6,6 is tough and rigid and does not need to be lubricated. It has a relatively high use temperature (to about 520 °F or 270 °C) and is used in the manufacture of items ranging from automotive gears to hair combs.

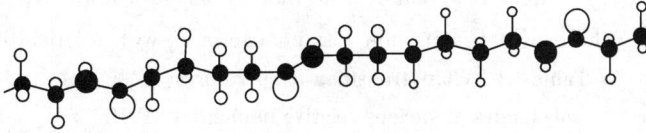

Figure 3.1   Ball-and-stick model of nylon 6,6

Most polymers progress from a glass solid to a softer solid and then to a viscous "taffy-like" stage allowing easy heat-associated fabrication. Nylon 6,6 has an unusually sharp transition from the solid to the soft stage requiring that fabrication be closely watched.

Nylon 4,6 was developed by DSM Engineering Plastics in 1990 and sold under the trade name Stanyl, giving a nylon that has a higher heat and chemical resistance for use in the automotive industry and in electrical applications. It has a T_m of 560 °F (295 ℃) and can be made more crystalline than nylon 6,6.

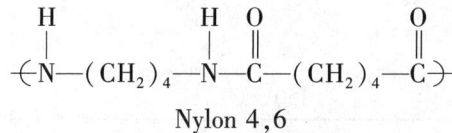

Nylon 4,6

A number of aromatic polyamides, aramids, have been produced that are strong, can operate under high temperatures, and have good flame-retardant properties. Nomex[TM] is used in flame-resistant clothing and in the form of thin pads to protect sintered silica-fiber mats from stress and vibrations during the flight of the space shuttles. Kevlar[TM] is structurally similar, and by weight it has a higher strength and modulus than steel and is used in the manufacture of so-called bullet resistant clothing. Because of its outstanding strength/weight ratio, it was used as the skin covering of the Gossamer Albatross, which was flown over the English Channel using only human power.

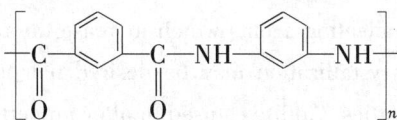
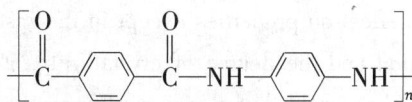

Nomex   Kevlar

Water absorption is characteristic of nylons. Unless compensated for by increased crystallinity, a higher proportion of amide groups leads to higher water adsorption. Increased water content has an effect analogous to that of increased temperature, i.e., enhanced segmental mobility with, for example, concomitant loss in stiffness and tensile strength, gain in toughness, and growth in dimensions (elongation). At very low temperatures, however, water stiffens the nylon. Thus, the brittleness temperature (ASTM D 746) of PA 66 is −80 ℃ if dry and −65 ℃ if conditioned to 50% relative humidity. Properties are frequently reported in the "dry", as-molded condition corresponding to about 0.2% water or less, and after equilibration to a specified relative humidity such as 50% or 65%, and occasionally to 100%. The greatest change occurs in the vicinity of the glass transition temperature (T_g) so that a useful aid in understanding behavior is a knowledge of the effect of humidity on T_g. This is complicated because of variation with the method of calculation or measurement (Table 3.1), but it is generally true that nylons with fewer CONH groups and lower water adsorption have a lower dry T_g but show less change of T_g with relative humidity.

Table 3.1 Glass transition temperatures (T_g, ℃) of polyamides at various relative humidities (R.H.) *

PA	Dry				50%R.H.		100%R.H.	
	A	B	C	D	C	D	C	D
46	102	78	78					
66	82	65	80,78,66	48	35	15	−15	−37
6(extracted)	56	65	75,65	41	20	3	−22	−32
610	56	46	67,70	42	40	10		
612	52	40	60	45	40	20	20	
11	36	29	53	43				
12	29	25	54	42			42	
MXD6	71		68					
TMDT			150					

A) Estimated from melting temperatures using $T_g = (2/3) T_m$;

B) Calculated from group contributions;

C) Dynamic measurements with torsion pendulum;

D) Static measurements by differential thermal analysis or inflection point in curve of modulus of elasticity as function of temperature.

Additives are materials used in small amounts (less than 5 wt% and usually less than 1 wt%) to affect processing, properties, or appearance. Processing additives include coloration inhibitors, lubricants, mold-release agents, nucleating agents, and viscosity thickeners or reducers. They have little or no effect on properties except in the instance of nucleating agents which increase the rate of crystallization and the degree of crystallinity. Enhanced crystallization may be desired not only to shorten molding cycles but also to modify mechanical properties. Additives used to alter properties or appearance are antioxidants, antistatic agents, biodegradative agents, biopreservatives, blowing agents, colorants, fragrances, and stabilizers against hydrolysis, thermal degradation, or UV degradation.

Modifiers are used in larger amounts, usually more than 5 wt %, in order to achieve desired changes in properties. Examples are mineral fillers, glass or carbon fibers, lubricants to improve wear and friction, plasticizers, fire retardants, electrically conductive materials, and other polymers to toughen or otherwise affect the nylon.

Alternative techniques of modification include copolymerization, adjusting molecular mass and post-treatments such as annealing, conditioning to some moisture level, dyeing, metallizing, painting, irradiation, or chemical reaction, e.g., alkoxyalkylation.

The emphasis here is on the more important modifiers and alternative techniques. Selection is necessary in light of the fact that the industry offers over 1 500 compositions (grades); that number includes the inevitable duplications from different suppliers.

Nylons are used in many and diverse ways. They are found in appliances, business equipment, consumer products, electrical/electronic devices, furniture, hardware, machinery, packaging, and transportation. This diversity makes classification and analysis difficult as shown in Table 3.2 which compares the pattern of consumption in the United States with that in Western Europe.

Table 3.2   **Pattern of nylon consumption in the United States and Western Europe in 1988**

Class	Percentage use in	
	United States	Western Europe
Appliances and power tools	3.0	
Building		6
Consumer products	6.1	
Electrical and electronic applications	7.7	20
Export	13.7	
Extrusion markets	25.8	23
Furniture/household		5
Industrial equipment	7.7	
Machinery		8
Miscellaneous	9.6	6
Transportation	26.4	32
Total percent	100.0	100
Total production. t/a	261 000	320 000

Transportation is the largest market for nylons. Unreinforced resins are used in electrical connectors, wire jackets, windshield wiper and speedometer gears, and emission canisters. The softer nylons are used in fuel lines, air brake hoses, and spline shaft coatings. Glass reinforced nylons are found in engine fans, radiator headers and grilles, brake and power steering fluid reservoirs, valve covers, wheel caps, air brake contacts and head-rest shells. Applications for nylons combining tougheners and reinforcement include brackets, steering wheels, and accelerator and clutch pedals. Mineral-filled resins are used in wheel caps, radiator grilles, and mirror housings. Nylons containing both glass fibers and minerals are used in exterior parts such as fender extensions. Toughened nylons are found in stone shields and trim clips.

Electrical and electronic applications comprise a major market for nylons, albeit more so in Western Europe than in the United States. Flame-retardant materials are particularly important in this area. Uses include color-coded components, plugs, connectors, coil forms, wiring devices, terminal blocks, antenna mounting devices, and harness ties. Wire and cable jacketing is used mostly over primary insulation because of the solvent, wear and abrasion resistance of nylons. Relays, fittings, and contact makers constitute a partial list of applications in telecommunications.

Industrial applications are attracted to the excellent fatigue resistance and repeated impact strength of nylons. Examples are hammer handles and moving machine parts. The mechanical strength accounts for use in gears, bearings, antifriction parts, snap fits, and detents. Food- and textile-processing equipment, pumps, valves, agricultural and printing devices, and business and vending machines comprise a partial list of other industrial uses.

Consumer products exploit the toughness of nylons in ski boots, ice and roller skate supports, racquet equipment, and bicycle wheels. Kitchen utensils, toys, photographic equipment, brush bristles, fishing line, sewing thread, and lawn and garden equipment show that the breadth of utility of nylon also includes consumer products.

Appliances and power tools make use of the impact strength of nylons. Glass-rein-forced resins, which combine stiffness at high temperatures with toughness and grease resistance, are used in handles, housings, and parts in contact with hot metal. Sewing machines, laundry equipment and dishwashers are examples of other appliances that have utilized nylons.

Film applications have grown in importance because of the use of nylons in co-extruded, composite films for food packaging. It is used also in cook-in bags and pouches. A relatively recent development involves dispersion of nylon platelets in a polyolefin matrix to combine both hydrocarbon and moisture barrier properties in blow molded containers.

The ways Nylon tailored to market needs are shown in Figure 3.2.

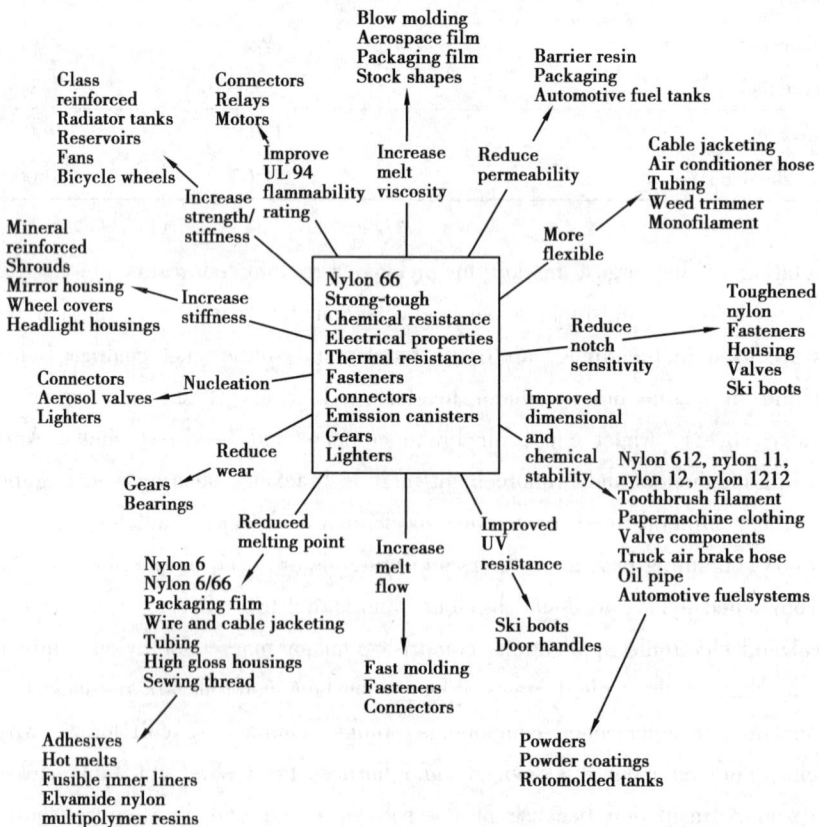

Figure 3.2   Nylon readily tailored to market needs

3.3　Polyesters

The history of thermoplastic polyesters goes back to 1929 with the pioneering work of carothers. The first thermoplastically processible polyesters synthesized from adipic acid and ethylene glycol were described by him in 1932. Polyesters only became of industrial interest in 1941, with the synthesis of high melting point products based on terephthalic acid[2].

The rapid industrial development of polyesters after World War II was initially restricted to polyester fibers based on poly (ethylene terephthalate) (PETP), poly (oxy-1, 2-ethanediyloxycarbonyl-1, 4-phenylenecarbony1). A polyester based on poly(1,4-dimethylenecyclohexane terephthalate) (PDCT), occupies a special position.

Poly (ethylene terephthalate) was subsequently also used in the production of films. Thermoplastic polyesters were first employed as construction materials in 1966, and were initially based on poly (ethylene terephthalate). In 1970 the more readily processible poly (butylene terephthalate) (PBT), poly(tetramethylene terephthalate), poly(oxy-1,4-butanediyloxycarbonyl-1,4-phenylenecarbonyl) was introduced into the market.

A short time later these homopolyesters were supplemented by a series of thermoplastic copolyesters which are suitable for specific areas of application due to their special properties. These copolyesters include polyetherester block copolymers as thermoplastic elastomers; copolyesters based on 1, 4-cyclohexanedimethanol for hard, glass-clear injection molded articles; and other thermoplastic copolyesters of varying composition for powder coatings, paint binders and hot-melt adhesives.

Recently a new class of fully aromatic thermoplastic polyesters has been developed-liquid crystalline polyesters. These polymers have aroused great interest on account of their outstanding mechanical and thermal properties. Two such products are commercially available, one being synthesized from 4-hydroxybenzoic acid and 6-hydroxy-2-naphthoic acid, and the other from 4-hydroxybenzoic acid, terephthalic acid, and 4,4'-dihydroxybiphenyl.

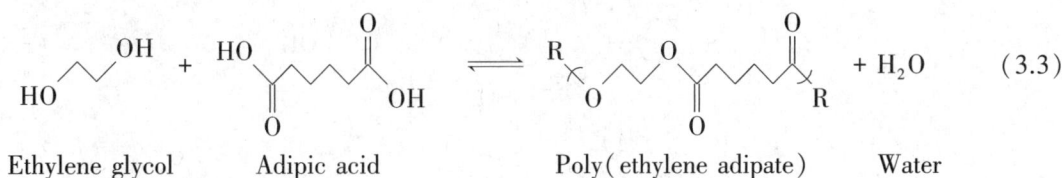

$$\qquad\qquad\qquad\qquad\qquad\qquad\qquad\qquad\qquad\qquad\qquad\qquad (3.3)$$

| Ethylene glycol | Adipic acid | Poly(ethylene adipate) | Water |

The DuPont research turned from the synthesis of polyesters to tackle, more successfully, the synthesis of the first synthetic fiber material, nylon, that approached, and in some cases exceeded, the physical properties of natural analogues. The initial experience with polyesters was put to use in the nylon venture.

The initial polyester formation actually occurred much earlier and is attributed to Gay Lussac and Pelouze in 1833 and Berzelius in 1847. These polyesters are called glyptals and alkyds, and they are useful as coatings materials and not for fiber production. While these reactions had low fractional conversions, they formed high-molecular-weight materials because they had functionalities (that is, number of reactive groups on a single reactant) greater than two, resulting in crosslinking.

The heat resistance of Carothers' polyesters was not sufficient to withstand the temperature of the hot ironing process. Expanding on the work of Carothers and Hill on polyesters, Whinfield and Dickson, in England, overcame the problems of Carothers and co-workers by employing an ester interchange reaction between ethylene glycol and the methyl ester of terephthalic acid, forming the polyester poly(ethylene terephthalate), PET, with the first plant coming on line in 1953. Plastic bottle is one of the famous products in polyester (Figure 3.3). This classic reaction producing Dacron, Kodel, and Terylene fibers and Dacron fibers is shown in Equation 3.4.

Figure 3.3   Molded PET bottle

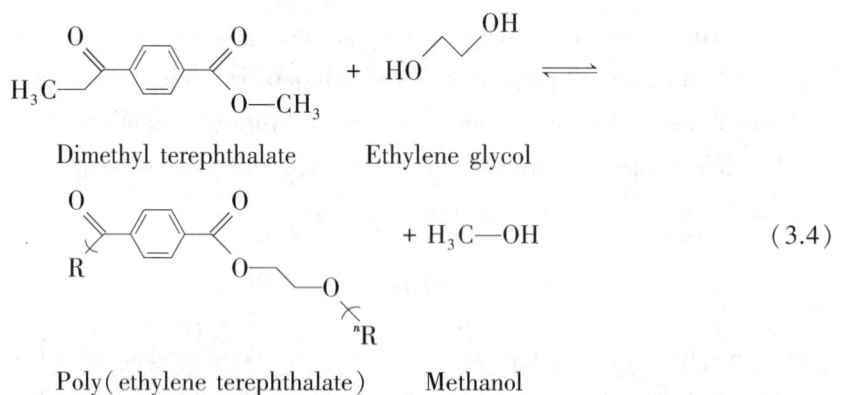

Dimethyl terephthalate   Ethylene glycol

$$+ H_3C-OH \qquad (3.4)$$

Poly(ethylene terephthalate)   Methanol

Methyl alcohol, methanol, is lower boiling than water (65 ℃ compared with 100 ℃) and thus more easily removed, allowing the reaction to be forced toward polymer formation more easily. This illustrates how similar materials can be made from more than one chemical reaction. While the poly (aryl esters), now simply called polyesters, produced by Whinfield and Dickson met most of the specifications for a useful synthetic fiber but because of inferior molding machines and inadequate plastic technology, it was not possible to injection mold these materials until more recently.

PBT has a melting point around 224 ℃ and is generally processed at about 250 ℃, whereas PET has a melting point around 270 ℃. This is a direct consequence of the presence of the addition of two methylene units in PBT, which allows easier fabrication of PBT by injection molding and extrusion procedures and through blow molding. PBT is employed for under the hood automotive parts, including fuse cables, pump housings, and electrical connectors, and for selected automotive exterior parts.

A more hydrophobic, stiffer polyester was introduced in 1958 by Eastman-Kodak as Kodel polyester. It contains a cyclohexanedimethanol moiety in place of the simple methylene moiety present in PET and PBT. Copolyesters based on cyclohexanedimethanol and ethylene glycol as the diols are blow-molded into bottles for shampoos and liquid detergents. Other, similar copolyesters are processed by extrusion into tough, clear fibers employed to package hardware and other heavy items. The comparing of unfilled and reinforced PET are shown in Table 3.3.

Table 3.3   Comparative data for poly(butylene terephthalate) and reinforced PBT

Property	Unfilled	30% Glass Reinforced
Tensile strength (psi)	8 000	18 000
Elongation (%)	150	3
Compressive strength (psi)	12 000	21 000
Flexural strength (psi)	14 000	27 000
Flexural modulus (psi)	350 000	1 100 000
Notched Izod impact (ft-lb/in. of notch)	1	1.5
Coefficient of expansion (cm/cm, ℃) (10)-6	75	25
Heat deflection temperature (℃)	65	220

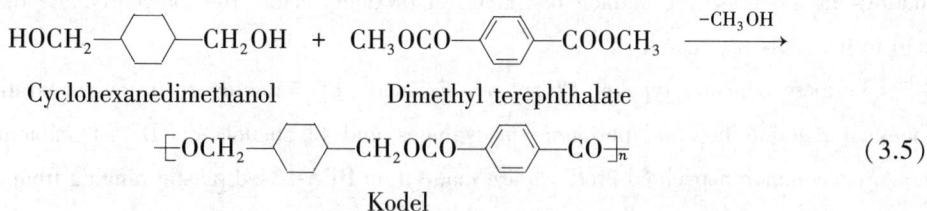

$$HOCH_2-\langle \rangle-CH_2OH \; + \; CH_3OCO-\langle \rangle-COOCH_3 \xrightarrow{-CH_3OH}$$

Cyclohexanedimethanol   Dimethyl terephthalate

$$\left[OCH_2-\langle \rangle-CH_2OCO-\langle \rangle-CO\right]_n \qquad (3.5)$$

Kodel

Several "wholly" aromatic polyesters are available. As expected, they are more difficult to process and are stiffer and less soluble, but they are employed because of their good high thermal performance. Ekonol is the homopolymer formed from ρ-hydroxybenzoic acid (below). Ekonol has a T_g in excess of 930 °F (500 ℃). It is highly crystalline and offers good strength.

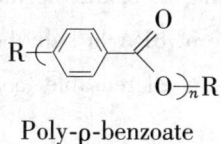

$$R\left[\langle \rangle \overset{O}{\underset{O}{\overset{\|}{C}}}\right]_n R$$

Poly-ρ-benzoate

It is not unexpected that such aromatic polyesters have properties similar to those of polycarbonates because of their structural similarities.

3.4 Polycarbonates[3]

Polycarbonate resin (PC resin) is a type of thermoplastic component used in the manufacturing of certain plastics. While the specific attributes of each resin differs according to its exact composition and extraction method, each is synthesized via catalyst from monomers called hydrocarbons, in a process known as condensation polymerization. The monomers used to produce polycarbonate resin differ from other types in that they contain amino, alcohol, or carboxylic acid functional groups. The chain reaction results in a covalent bond of one carbon atom bonded to three oxygen atoms, with small water molecules being displaced to yield a final polymer of high density and impact strength. These properties make polycarbonate resin suitable for manufacturing a wide variety of products that require exceptional stress and heat resistance, such as compact discs, hockey masks, eyeglasses, automobile parts, bulletproof glass, medical and aerospace equipment, and even shuttle parts for the U.S. National Aeronautics and Space Administration (NASA) space program.

Many products like those mentioned above are made by subjecting a polycarbonate resin to an injection molding process in which the polymerized material hardens inside a mold or die and permanently takes on its parent shape. In some cases, the end product may require treatment with a coating to provide additional insurance against damage from chronic exposure to ultraviolet radiation or certain chemicals. For instance, PC plastics cannot withstand contact with solvents such as benzene, acetone, or sodium hypochlorite, otherwise known as household bleach. Protective coatings also increase the surface resistance of products made from polycarbonate resin since they tend to be easily scratched.

The most common type of PC plastic is made by inducing resin polymerization through a chemical reaction between phosgene isocyanates and bisphenol A (BPA) monomers. In fact, numerous common household products are made from BPA-based plastic ranging from electronic and computer components to baby bottles and food storage containers. Since BPA is now known to be an endocrine disruptor associated with infertility, birth defects, neurological disorders, and hormone-dependent cancers, its impact on human health for more than half a century is of great concern, as well as its continued use. Yet, in spite of numerous international studies and reports on the subject, very few countries have considered banning or modifying the use of BPA in the plastics industry. However, Denmark moved to ban the use of BPA in baby bottles in 2009, and several U.S. states have independently banned its inclusion in all reusable food and beverage containers, as well as those that store infant formula and food.

Einhorn produced a high-melting, clear polyester of carbonic acid in 1898 by the reaction of phosgene and hydroquinone or resorcinol. Commercial polycarbonates (PC) were produced in the 1950s by the General Electric Company and Bayer Company by the condensation of bisphenol A and phosgene ($COCl_2$). This tough, transparent polymer (Lexan, Merlon) is produced at an annual rate of 130 000 tons.

Polycarbonates are processed by all the standard plastic methods. They are used in glazing (40%), appliances (15%), signs, returnable bottles, solar collectors, business machines, and electronics. They show good creep resistance, good thermal stability, and a wide range of use temperatures ($-50\ ℃ \sim 130\ ℃$). Coatings are generally used on PC sheets to improve mar and chemical resistance.

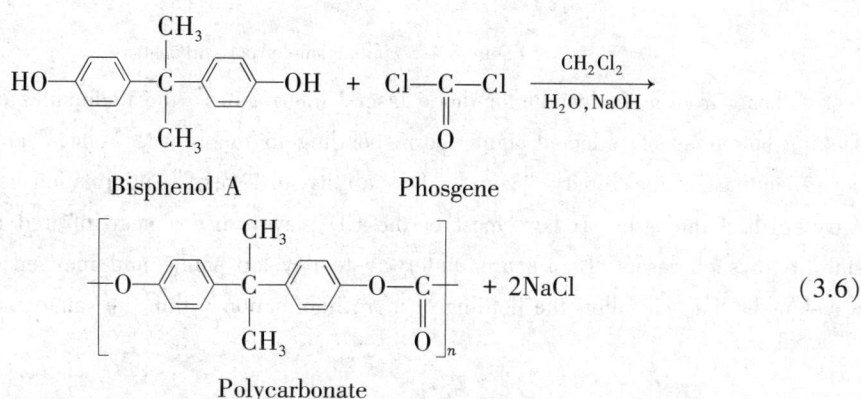

$$HO-\!\!\!\bigcirc\!\!\!-\!\!\overset{\overset{\displaystyle CH_3}{|}}{\underset{\underset{\displaystyle CH_3}{|}}{C}}\!\!-\!\!\bigcirc\!\!\!-OH \;+\; Cl-\overset{}{\underset{\underset{\displaystyle O}{\|}}{C}}-Cl \;\xrightarrow[H_2O,\,NaOH]{CH_2Cl_2}$$

Bisphenol A　　　　　　　Phosgene

$$\left[\!\!-O-\!\!\bigcirc\!\!\!-\!\!\overset{\overset{\displaystyle CH_3}{|}}{\underset{\underset{\displaystyle CH_3}{|}}{C}}\!\!-\!\!\bigcirc\!\!\!-O-\overset{}{\underset{\underset{\displaystyle O}{\|}}{C}}\!\!-\!\!\right]_n \;+\; 2NaCl \qquad (3.6)$$

Polycarbonate

Blends of PC and ABS (Bayblend) or with poly(butylene terephthalate) (Xenoy) are tough, heat-resistant (HDT 200 ℃) plastics. The properties of polycarbonates are shown in Table 3.4.

Table 3.4　**Comparative data for polycarbonate and reinforced polycarbonate**

(**Lexan, Merlon**)

Property	Unfilled	30% Glass Reinforced
Tensile strength (psi)	9 500	19 000
Elongation (%)	110	4
Compressive strength (psi)	12 500	18 000
Flexural strength (psi)	13 500	23 000
Flexural modulus (psi)	340 000	11 000 000
Notched Izod impact (ft-lb/in. of notch)	14	2
Coefficient of expansion (cm/cm, ℃) (10)-6	68	22
Heat deflection temperature (℃)	130	146

Nonrecordable compact discs (CDs) are made of rigid, transparent polycarbonates with a reflective metal coating on top of the polycarbonate. A laser is used to encode information through creation of physical features sometimes referred to as "pits and lands" of different reflectivity at the polycarbonate-metal interface.

Recordable CDs contain an organic dye between the polycarbonate and metal film (Figure 3.4). Here, a laser creates areas of differing reflectiveness in the dye layer through photochemical reactions.

(a)   (b)

Figure 3.4   Applications of polycarbonates

A beam from a semiconductor diode laser "interrogates" the undersides of both types of CDs seeking out areas of reflected light, corresponding to the binary "one," and unreflected light, corresponding to the binary "zero." The ability to "read" information is dependent on the wavelength of the laser. Today, most of the CD players use a near-infrared laser because of the stability of such lasers. Efforts are underway to develop stable and inexpensive lasers of shorter wavelengths that will allow the holding of more information within the same space.

3.5   Polyacetals[4]

Acetal polymers, also known as polyoxymethylene (POM) or polyacetal are formaldehyde-based thermoplastics that have been commercially available for 40 years. Polyformaldehyde itself (i.e., the homopolymer of polyacetal) is a thermally unstable material that decomposes on heating to yield formaldehyde gas. Two methods of stabilizing polyformaldehyde for use as an engineering polymer were developed and introduced by DuPont in 1959 and Celanese in 1962.

DuPont's route for polyacetal yields a homopolymer through the esterification of polyformaldehyde with acetic acid to give a polyformaldehyde-acetic anhydride copolymer.

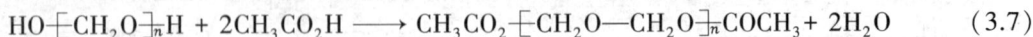

$$HO\text{--}[CH_2O]_n H + 2CH_3CO_2H \longrightarrow CH_3CO_2\text{--}[CH_2O\text{--}CH_2O]_n COCH_3 + 2H_2O \qquad (3.7)$$

The Celanese route for the production of polyacetal yields a more stable copolymer product via the reaction of trioxane, a cyclic trimer of formaldehyde, and a cyclic ether (e.g., ethylene oxide as shown below; or 1,3-dioxolane-$(CH_2)_2O_2CH_2$):

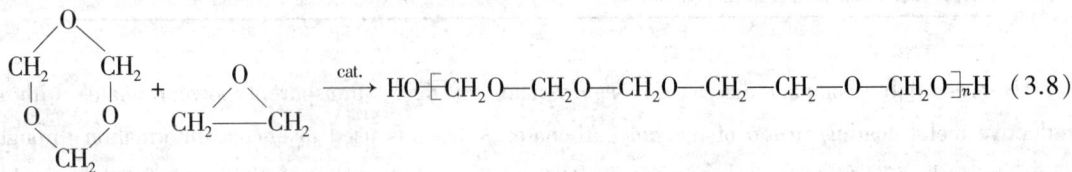

$$\begin{array}{c} \text{trioxane} \end{array} + \begin{array}{c} \text{ethylene oxide} \end{array} \xrightarrow{\text{cat.}} HO\text{--}[CH_2O\text{--}CH_2O\text{--}CH_2O\text{--}CH_2\text{--}CH_2\text{--}O\text{--}CH_2O]_n H \qquad (3.8)$$

Oxyethylene Linkage

The improved thermal and chemical stability of the copolymer versus the homopolymer is a result of randomly distributed oxyethylene groups. These groups offer stability to oxidative, thermal, and alkaline attack. The raw copolymer is hydrolyzed to an oxyethylene end cap to provide thermally stable polyacetal copolymer.

Polyacetal homopolymer (for example, DELRIN, produced commercially by DuPont) is a polyoxymethylene material produced by polymerization of formaldehyde. The formaldehyde, which is supplied to the polymerizer as a vapor, must have a purity of greater than 99.9 weight percent in order to achieve polymer molecular weights in the commercially desired range (20 000 or greater). Water, methanol, and formic acid, present in commercial formaldehyde, act as chain-transfer agents that lower the polymer molecular weight and must therefore be removed. Conceptual process flow diagrams are provided.

Due to its improved thermal stability, the copolymer has become the preferred form. The commercially significant polyacetal copolymers (for example, CELCON, produced by Ticona (the plastics division of Celanese Corporation)) are addition polymers of formaldehyde and a relatively small amount of a comonomer.

Polyacetal copolymer is produced by a bulk polymerization process. A strong Lewis acid serves as the polymerization catalyst and a chain-transfer agent is added to control molecular weight. Boron trifluoride is the Lewis acid most commonly used today as a catalyst. It is important to remove any water or methanol, as water yields no stable end-groups, and methanol yields only one stable end-group. Hence, as in the homopolymer process, it is important that concentration of water in the streams supplied to the reactor be held to a minimum. Conceptual process flow diagrams are provided.

It is widely recognized that the copolymer processes currently in commercial operation include additional steps in order to deactivate the catalyst and stabilize the acetal resin. Recent patents discuss a simplified process that significantly reduces the capital costs and operating steps required to produce polyacetal resins. Simply stated, the process uses a strong protic acid to catalyze the polymerization reaction, instead of a Lewis acid. The copolymer acetal resin is then transferred directly from the polymerization reactor directly into a twin screw extruder for finishing. No deactivation steps are required, and removal of formaldehyde or excess monomer is accomplished by drawing a vacuum on the material as it is transferred and as it is being finished.

It does not appear that such a simplified process will work in the production of homopolymer acetal resins, as there are no copolymer units to block depolymerization of the polymer chains. Consequently, homopolymer polyacetal producers have worked to improve the throughput of their processes, most likely by improving the on-stream factors for their polymerizers by reducing polymer fouling of the top heads and vapor lines.

Polyacetals are employed in plumbing and irrigation because they resist scale accumulation and have good thread strength, torque retention, and creep resistance. Polyacetals are used for molded door handles, tea kettles, pump impellers, shoe heels, and plumbing fixtures. Properties of these polymers are shown in Table 3.5.

Table 3.5 **Properties of high-performance plastics**

Property	ASTM Method	Acetal Copolymer	Acetal Homopolymer
Specific gravity (g/cm^3)	D792	1.410	1.425
Tensile strength at yield (psi)	D638	8 800	10 000
Elongation at break (%)	D638	60	25
Tensile modulus (psi, ×10^5)	D638	4.10	5.20
Flexural strength (psi,)	D790	13 000	14 100
Flexural modulus (psi, ×10^5)	D790	3.75	4.10
Fatigue endurance limit (psi/no. of cycles)	D671	4 200 per	5 000 per
Compressive stress at 10% dilation (psi)	D695	16 000	18 000
Rockwell hardness (M)	D785	80	94
Notched Izod impact (ft-lb/in. of notch)	D256	1.3	1.4
Tensile impact (ft-lb/in.^{2})	D1822	70	94
Water absorption, 24 h immersion (%)	D570	0.22	0.26
Tabor abrasion, 1 000-g load, Cs-17 wheel(mp/1 000 cycles)	D1044	14	20

Polyacetal is only one of a number of engineering thermoplastics (ETPs), as illustrated in Figure 3.5.

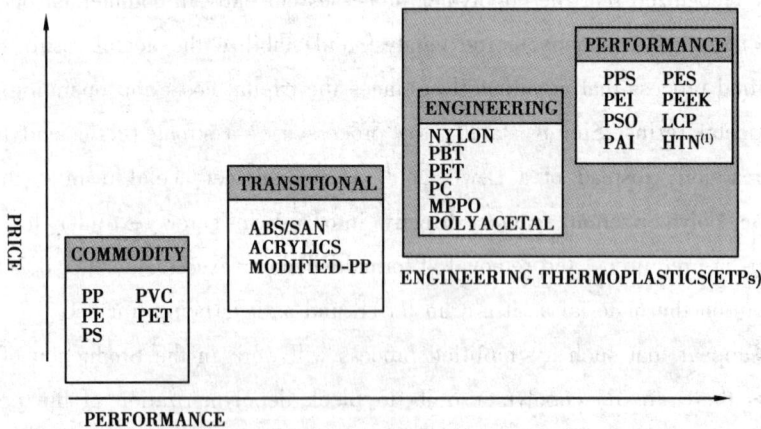

Figure 3.5 Price-performance comparison of thermoplastics

Although acetal resins compete quite strongly against nylon 6 and 6,6, they do, at times, compete against other materials in the Engineering Thermoplastic family. This is partly a result of the versatility of these materials and partly a result of property deficiencies being overcome by blending, additives, or other techniques.

Polyacetal's versatility and usefulness in small parts is largely responsible for the demand by end-use pattern shown in Figure 3.6, where the largest end-use sector is electrical/electronics (E/E) and appliances at 31 percent of the global total. Note that demand is fairly evenly spread among the next three largest end-use markets, a pattern which is unusual in engineering plastics.

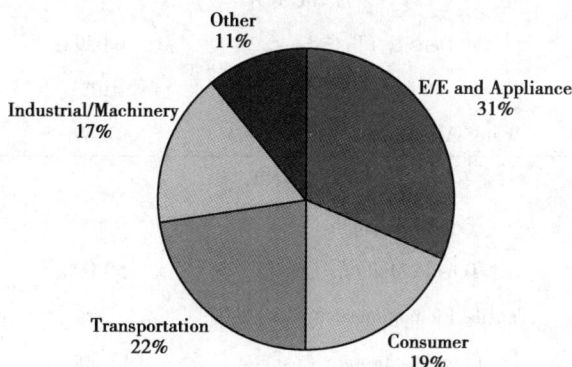

Figure 3.6   Global polyacetal demand by end-use

3.6   Poly(phenylene oxide)

In 1956, A. Hay of the GE discovered a convenient catalytic oxidative coupling route to high molecular weight aromatic ethers. Polymers were made by bubbling oxygen through a copper-amine-catalyzed solution of phenolic monomer at room temperature[6]. A wide variety of phenolic compounds were explored, but the cleanest reactions resulted from those that contained small, electron-donor substituents in the two ortho positions. Hence, research quickly focused on 2,6-dimethylphenol and poly(2,6-dimethyl-1,4-phenylene ether) or PPO. The synthesis is shown below.

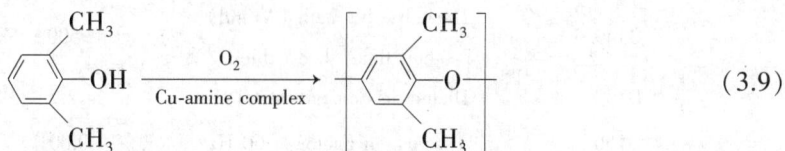

$$\text{(3.9)}$$

In 1964 GE introduced the PPO homopolymer. PPO had excellent hydrolytic resistance and an extremely high T_g of 215 ℃. However, the very high melt viscosity and a pronounced tendency to oxidize at processing temperatures made PPO very difficult to process. And then, GE introduced a family of PPO blends with high impact polystyrene (HIPS), under the Noryl trademark. This combination of total compatibility with PS and the added toughness from HIPS was the key to commercial success. Varying the PPO/HIPS ratio results in a wide range of high temperature, easy

79

to process, tough and dimensionally stable plastics[7]. The properties of modified PPO are shown in Table 3.6.

Table 3.6 Typical Properties of NORYL®(Modified PPO)[5]

ASTM or UL test	Property	Unfilled	30% Glass-filled
	PHYSICAL		
D792	Density (lb/in^3)	0.039	0.049
	(g/cm^3)	1.08	1.36
D570	Water Absorption, 24 hrs (%)	0.07	0.06
	MECHANICAL		
D638	Tensile Strength (psi)	9 600	17 800
D638	Tensile Modulus (psi)	350 000	—
D638	Tensile Elongation at Break (%)	30	—
D790	Flexural Strength (psi)	13 500	20 000
D790	Flexural Modulus (psi)	360 000	1 100 000
D785	Hardness, Rockwell	R119	L108
D256	IZOD Notched Impact (ft-lb/in)	5.0	2.3
	THERMAL		
D696	Coefficient of Linear Thermal Expansion(10^{-5}in./in./°F)	3.3	1.4
D648	Heat Deflection Temp (°F/℃) at 264 psi	265/129	275/135
D3418	Vicat Softening Temp (°F/℃)	310/154	—/—
—	Max Operating Temp (°F/℃)	220/105	220/105
C177	Thermal Conductivity (10^{-4} cal/cm-sec-°C)	4.55	6.68
UL94	Flammability Rating (1/4″ thick)	V-0	V-0
	ELECTRICAL		
D149	Dielectric Strength (V/mil) short time, 1/8" thick	500	530
D150	Dielectric Constant at 60 Hz	2.7	3.2
D150	Dissipation Factor at 60 Hz	0.001	0.002
D257	Volume Resistivity (Ω) at 50% RH	10^{17}	10^{17}

Noryl resins became the world's most successful and best-known polymer blends because combinations of PPO resins and styrene polymers tend to assume the best features of each:

①PPO resins with very high heat distortion temperatures (HDTs) can readily raise the HDT of styrenics to over 100 ℃, which is a significant temperature because this qualifies the product for all boiling water applications.

②HIPS, with ease of processing and well-established impact modification, balances the refractory nature of PPO resins.

③PPO resins bring fire retardance to the system.

④Both PPO resins and styrene polymers have excellent water resistance and outstanding electrical properties.

⑤In addition, PPO/PS blends exhibit lower specific gravity than other engineering thermoplastics.

The first applications were those requiring autoclaving (medical equipment) and outstanding electrical properties at elevated temperatures. As compounding, stabilization, and processing skills improved, markets for PPO blends expanded to include office equipment, electronic components, automotive parts, water distribution systems, and general metal replacement.

PPO-based resins are relatively resistant to burning, and judicious compounding can increase their burn resistance without the use of halogenated flame retardants. They may be modified with glass and other mineral fillers. Because of low moisture absorption, dimensional stability, and ability to be used over a wide temperature range, PPO-based resins are especially adaptable to metallization. After expiration of the original patent on PPO, several manufacturers (e.g., BASF, Huls, Borg-Warner, Asahi, Mitsubishi) began sales of their own blends based on PPO or its copolymers.

3.7   Poly(aryl sulfones)

Polyaryl sulfone (PSU) is a heat-resistant, transparent, amber, non-crystalline engineering plastic having the molecular structure of diphenylene sulfone group. The influence of diphenylene sulfone on the properties of resins has been the subject of intense investigation since the early 1960s. The contributions of this group become evident upon examination of its electronic characteristics. The sulfur atom (in each group) is in its highest state of oxidation. Furthermore, the sulfone group tends to draw electrons from the adjacent benzene rings, making them electron-deficient. Thermal stability is also provided by the highly resonant structure of the diphenylene sulfone group. This high degree of resonance imparts high strength to the chemical bonds. Substances stable to oxidation strongly resist the tendency to lose their electrons to an oxidizer. It then follows that the entire diphenylene sulfone group is inherently resistant to oxidation[8].

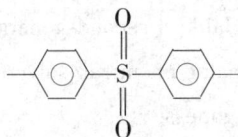

Diphenylene sulfone

PSU includes polyether sulfone (PES), polyphenyl sulfone (PPSU) etc..PES is a tough and rigid resin similar to conventional engineering plastics, such as polycarbonate, at room temperature[9]. The greatest characteristic of PES is that it has by far better high-temperature properties than conventional engineering plastics. Specifically, PES remains in satisfactory condition in long-term continuous use without causing any dimensional change or physical deterioration at temperatures as high as 200 ℃. Table 3.7 lists some commercially available polysulfones.

Table 3.7 Commercially available polysulfones

Trade Name	Polymer Unit	$T_g/℃$
Astrel (3m Corp.)		285
Poly(ether sulfone) 720 P(ICI)		250
Poly(ether sulfone) 200 P(ICI)		230
Udel (Union Carbide)		190

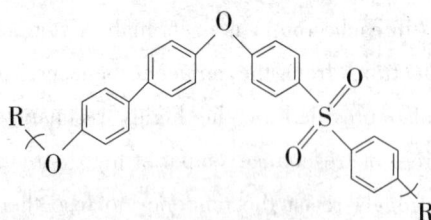

Radel

In 1976, Union Carbide made available a second-generation polysulfone under the trade name of Radel. This polysulfone exhibited greater chemical/solvent resistance, a greater T_g of 220 ℃, greater oxidative stability, and good toughness.

Complex-shaped objects can be made through injection molding without need for additional machining and other procedures. Films and foil are used for flexible printed circuitry. Polysulfones are also used for ignition components, hair dryers, cookware, and structural foams. Because of their good hydrolytic stability, good mechanical properties, and high thermal endurance they are good candidate materials for hot water and food handling equipment, alkaline battery cases, surgical and laboratory equipment, life support parts, autoclavable trays, tissue culture bottles, and surgical hollow shapes, and film for hot transparencies. Their low flammability and smoke production, again because of their tendency for polycyclic formation on thermolysis and presence of moieties that are partially oxidized, makes them useful as materials for aircraft and the automotive industry.

Table 3.8   **Typical properties of Radel**[10]

ASTM or UL test	Property	unfilled (A-300)	30% glass filled (AG-330)
	PHYSICAL		
D792	Density (lb/in^3)	0.050	0.057
	(g/cm^3)	1.37	1.58
D570	Water Absorption, 24 hrs (%)	0.54	0.39
	MECHANICAL		
D638	Tensile Strength (psi)	12 200	18 900
D638	Tensile Modulus (psi)	385 000	825 000
D638	Tensile Elongation at Yield (%)	6.5	1.9
D790	Flexural Strength (psi)	16 100	23 500
D790	Flexural Modulus (psi)	420 000	950 000
D695	Compressive Strength (psi)	14 500	25 600
D695	Compressive Modulus (psi)	388 00	1 119 000
D785	Hardness, Rockwell	M88/R127	M80/R124
D256	IZOD Notched Impact (ft-lb/in)	1.6	1.4
	THERMAL		
D696	Coefficient of Linear Thermal Expansion (10^{-5}in./in./°F)	2.7	1.7
D648	Heat Deflection Temp (°F/°C) at 264 psi	400/204	420/215
C177	Thermal Conductivity (10^{-4} cal/cm-sec-°C)	3.89	4.65
UL94	Flammability Rating	V0	V0
	ELECTRICAL		
D149	Dielectric Strength (V/mil) short time, 1/8" thick	380	440
D150	Dielectric Constant at 1 kHz	3.5	4.1
D150	Dissipation Factor at 1 kHz	0.002 2	0.001 8
D257	Volume Resistivity (Ω) at 50% RH	1.7×10^{15}	$> 10^{16}$

* RADEL and UDEL are registered trademarks of Solvay Advanced Polymers[11]

A typical example of a current innovation that only became possible thanks to the systematic development of the high-temperature amorphous thermoplastics is that of oil-regulating pistons (Figure 3.7). Ultrason KR 4113, a PESU from BASF, proved suitable for this component because of the dimensional stability conferred by its carbon fiber reinforcement. A crucial factor was that the material possesses outstanding tribological properties due to its 10% graphite and PTFE content. The potential offered by carbon-fiber-reinforced polyethersulfones, which include products such as Ultrason E 2010 C6 with 30 % fiber, is far from exhausted. In certain applications they can even compete with metals in terms of thermal expansion up to temperatures of around 200 ℃[12].

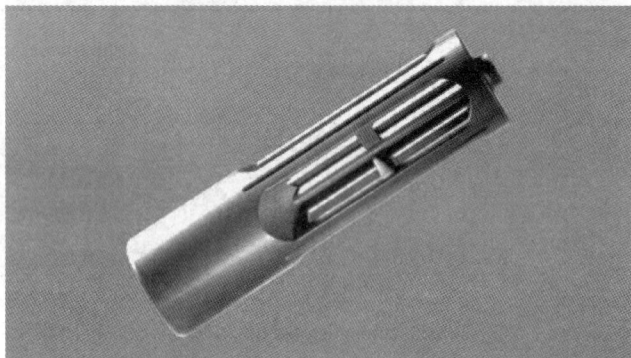

Figure 3.7 A polyethersulfone modified with graphite and PTFE offers
very good tribological properties for oil-regulating pistons

This example shows that polyarylsulfones, unlike other polymers, are not classic predatory polymers—they are growing strongly because of their inherent potential for innovation. PSU, PES, PPSU and PEI together make up about 15% of the market for high-temperature plastics; PEI accounts for about a third of this slice. World demand for these amorphous HT thermoplastics is currently running at around 50 kt, and the market is expected to exceed 100 kt by 2020. BASF alone, one of the largest providers of these HT thermoplastics (in addition to Solvay, Sumitomo and Sabic), doubled its Ultrason capacity at its Ludwigshafen site to 12 000 t/a at the end of 2007.

Headlight reflectors are still the main application for polysulfones in the automotive sector. The widespread halogen lamps develop a large amount of heat in use that the materials employed must withstand. In addition, the headlight ventilation system cannot remove the heat quickly. Measured reflector temperatures now often range from 180 to 200 ℃. Aside from high softening temperatures, the materials must possess very high dimensional stability specially given the current trend towards more complex designs—because the heat must not alter the shape and orientation of the light cone. The key advantage over thermoset BMC is the possibility of direct metallization. In this process, a thin layer of aluminum is deposited from a vacuum vapor or by vacuum sputtering onto freshly injection molded parts. This market has been well served for some time by polyethersulfones (PES), such as Ultrason E 2010 Q26 from BASF, and polyetherimides (PEI), which are supplied solely by Sabic. In 2006, Solvay, too, launched new, directly metallizable polyethersulfones for the lighting sector. They are designed for operating temperatures of up to 205 ℃ and are notable for their high melt flow rates (Radel LTG-3000 PES). In particular, it is difficult to realize new ideas such as cornering lights or highly stylized daytime running lights without PESU or PEI due to the high demands on the geometry. Free-flowing polysulfones, such as Udel LTG-2000 (manufacturer: Solvay), are also starting to compete with polycarbonate and (PC + PEI) blends in low-temperature areas. The amorphous HT thermoplastics are acting in direct competition here with thermosets (BMC). Despite a significantly higher material price, polyethersulfones have enjoyed marked growth in recent years due to advantages such as design freedom and weight savings: Even though the growth is not as fast as it was a few years ago, it is still running at 5% to 10% in the automotive sector and is thus far outstripping the automotive market itself.

The market for polysulfones in the electrical and electronics sector is also developing at a very lively rate, with this class of polymers enjoying steady growth rates of 4%. A major application area, in which there is still healthy demand for PSU and PES, is that of pawls for fuse boxes where the reduced creep and high dimensional stability of plastics comes to the fore. Products used here include the BASF polyethersulfones Ultrason E 2010 (of various glass fiber content) and E 3010 (particularly high impact and stress cracking resistance), and the polysulfone (PSU) Ultrason S 2010 G6. Otherwise, amorphous HT thermoplastics in the E&E sector are to be found in a variety of small applications, such as fuel cell components, where heat-resistance and dimensional stability are required. Further examples are housing parts of installations, such as air humidifiers and industrial batteries.

Among the innovations stimulated by polyethersulfones in electronics is the development of transparent displays (Figure 3.8). These are now made mostly from glass due to the high temperatures of up to 360 ℃ which arise during ITO coating (Indium tin oxide, a transparent semiconductor). Similar processes, however, can also be performed at temperatures which extend downwards into the domain of HT engineering plastics (glass transition temperatures above 220 ℃. The plastic displays are not only more

Figure 3.8   Polyethersulfones, such as Ultrason E 2010 HC, could well be facing an interesting future in lightweight, thin and flexible displays.

flexible, but in comparison to glass, much thinner. Here, there could be substantial potential for polyethersulfones such as Ultrason E 2010 HC (HC: high clarity, BASF) with their good transparency in the optical spectrum, although the technical requirements are not low. High demands have to be imposed on the materials' purity and freedom from nibs, which are important not only for optical quality, but also for achieving the required surface roughness values of less than 10 nm. The development of these displays is currently being forced in Asia, which explains the strong presence of Sumitomo, an Asian supplier, in this field.

Figure 3.9 In microwave dishes, polysulfones withstand the high temperatures of boiling grease, are not sensitive to hydrolysis and are very tough.

The household sector would find it hard to replace polyarylsulfones. PESU, PPSU and PSU are found in a wide range of applications here with and without fiber reinforcement and additives for, e.g., improving demolding behavior and UV stability. Fiber-reinforced variants, for example, are used for coffee-maker parts that come into contact with hot steam. Polysulfones are competing in this application against metal. It is no easy task to prevail in this sector. In addition to the high requirements imposed on food-contact plastics by the licensing authorities, the polymers must possess high hydrolytic resistance and low absorption of food colorants. In addition, they must be highly resistant to cleaning agents and the high temperatures which fats can reach in the microwave (Figure 3.9). At the same time, they must be very tough, which is important when plastic parts are cleaned alongside steel pots in industrial dishwashers. What is more, the materials must not fracture.

Another, less familiar role played by the polyethersulfones Ultrason E 2020 P and 2020 P SR lies in coated pans (Figure 3.10). While PTFE does not adhere to metals, it adheres very well to PES, which is why polyethersulfones are used as coupling agents to provide a bond between the non-stick layer and steel in such utensils. PES is also gaining importance in baby bottles: Polyethersulfones such as Ultrason E 2010 are readily sterilized at high temperatures. Household applications for PPSU include pipe fittings. Their purpose here is to suppress the formation of micro-cracks from the outset. For this application, BASF has recently launched a particularly high-impact, stress-cracking-resistant grade of Ultrason P 3010, which also lends itself to the production of rugged trays and aircraft interior trim. Sales of polyarylsulfones in the household sector have enjoyed steady growth for years.

Figure 3.10 PES acts as a coupling agent between the non-stick layer of PTFE and the steel of the cooking utensil.

Totally different properties are required of polyarylsulfones in the field of membrane technology. Here, advantage is taken of the fact that polysulfone films and filaments precipitated from organic solvents with water exhibit extraordinarily reproducible porosity. PSU and PESU membranes can be found, for example, in micro-and ultrafiltration, which plays an important role in beverage processing and water treatment (Figure 3.11). Sterilizable polyarylsulfones are therefore taking market share from traditional membrane materials. In principle, however, polysulfones are helping to boost the sales figures of their producers and are growing faster than the medical technology market itself.

Figure 3.11　Sterilizable polyarylsulfones are finding increasing use, for example, as hollow fiber bundles in micro- and ultrafiltration.

Aviation traditionally has great demand for high-end plastics. Again, polyarylsulfones have established themselves because they are inherently flame-retardant without the need for additives. Unlike many other polymeric materials, in which flame smoke density and toxicity increase when they are rendered flame retardant with additives, polyarylsulfones inherently have good FST properties (Flame Smoke Toxicity). Their excellent fire-safety properties also qualify polyarylsulfones for use in many other, equally safety-critical means of transport, such as subways.

Interesting opportunities are opening up in the aviation sector not only for polyarylsulfones themselves but also for the lightweight, carbon fiber-reinforced epoxy-carbon prepregs. One way to lower their crack sensitivity and raise their impact strength is to modify the resins by adding polyethersulfones. BASF offers the Ultrason grades E 2020 P and E 2020 P SR (Micro) for this highly knowledge-intensive business. The material comes in flakes or powder form to facilitate processing in common solvents. Products with similar characteristics are also obtainable from Sumitomo and Solvay. Readily soluble powder PESU grades, such as Sumikaexcel 5003 P (Sumitomo) still feature prominently in paints, durable coatings and heat-resistant adhesives; the polymer, which exhibits remarkable adhesion to glass, ceramics and metals thanks to a large number of free hydroxyl groups, can survive hot air temperatures of up to 250 ℃.

Successful entry into new applications is a characteristic feature of polyarylsulfones and it can be assumed that this trend will persist. One interesting way to publicize the extraordinary properties of this polymer class even more proceeds via the designers who decide on the materials at an early stage in the product-development process. BASF, especially, subscribes to this idea. The objective is to draw the attention of innovative and creative industrial designers through unusual model projects, such as a compact, transparent PESU toaster, directly to the remarkable properties of these polymer materials without proceeding via data sheets and technical manuals. The concept is bearing first fruits: a handle-less kettle made of transparent Ultrason (Figure 3. 12) won a prestigious award at a conference on materials for product development, design and architecture that was held in Frankfurt. Germany, in 2007.

Figure 3.12   This transparent designer kettle made from Ultrason,
which doe snot require a handle since the material possesses
low thermal conductivity, won a design award.

3.8   Polyimides

Polyimides are a class of thermally stable polymers that are often based on stiff aromatic backbones. The chemistry of polyimides is in itself a vast area with a large variety of monomers available and several methodologies available for synthesis. However, there has been considerable debate on the various reaction mechanisms involved in different synthesis methods. This review however, covers only the important fundamentals regarding the polyimide synthesis. The focus in this review will rest only on "aromatic" polyimides as they constitute the major category of such materials. Secondly, the properties of polyimides can be dramatically altered by minor variations in the structure. The subtle variations in the structures of the dianhydride and diamine components have a tremendous effect on the properties of the final polyimide. This chapter reviews several such features that are important towards understanding these structure-property relationships. Specifically, the effects of changing the diamines, dianhydrides or the overall flexibility of the chain on the basic parameters like T_g and T_m are also examined[13].

The most widely practiced procedure in polyimide synthesis is the two-step poly(amic acid) process. It involves reacting a dianhydride and a diamine at ambient conditions in a dipolar aprotic solvent such as N, N-dimethylacetamide (DMAc) or N-methylpyrrolidone (NMP) to yield the corresponding poly(amic acid), which is then cyclized into the final polyimide. This process involving a soluble polymer precursor was pioneered by workers at Dupont in 1950s, and to this

88

day, continues to be the primary route by which most polyimides are made. Most polyimides are infusible and insoluble due to their planar aromatic and hetero-aromatic structures and thus usually need to be processed from the solvent route. This method provided the first such solvent based route to process these polyimides. The process also enabled the first polyimide of significant commercial importance-Kapton™, to enter the market. The process for most extensively developed Kapton™ polyimide utilizes the monomer pyromellitic dianhydride (PMDA) and 4,4'-oxydianiline (ODA) and is illustrated in Figure 3.13.

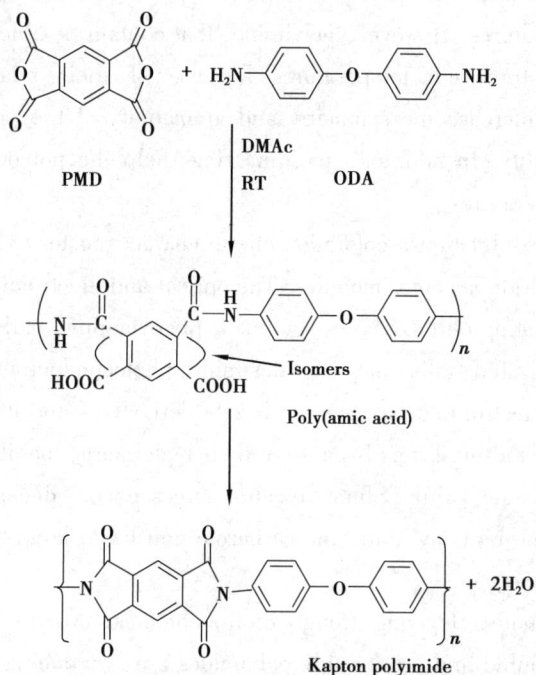

Figure 3.13   Reaction scheme for the preparation of Kapton™ polyimide

However, the seemingly simple process involves several elementary reactions that are interrelated in a complex scheme. The course of these reactions can be tremendously affected by a large number of factors that include reaction conditions and even the mode of monomer addition. The success of the overall reaction to yield high molecular weight polymers is critically dependent on seemingly subtle details. The ensuing discussion will address several parameters which govern these interrelations with respect to dependence of the synthesis on choice of monomers, solvents, reaction conditions and the importance of various side reactions involved in the synthesis.

The mechanical properties of polyimides are influenced by many factors, such as the chemical structure, viscosity, molecular weight, preparation procedure, heating history, sample preparation and the method of property determination. Therefore, there is no clear rule can be discerned for the mechanical properties of the polyimides. Thus it is likely that differences in mechanical properties are concealed by large experimental uncertainties. In general, polyimides exhibit modulus values of 1.5~3.0 GPa and tensile strengths of 70~100 MPa. However, the elongation at breakage ranges from 2% to 15%, depending on the chemical structure. Polyimides containing flexible linkage units, such as ether linkages and isopropylidene[14], in the main chain exhibit more elongation. In addition, noncoplannar, asymmetrical and amorphous polyimides also usually show higher elongation. It is a general rule but not absolute: the polymers with high mechanical modulus show lower elongation.

Thermogravimetric (TG) analysis reveals good thermal stability for aromatic polyimides. In general, polyimides are stable up to a temperature of 440 ℃ in a nitrogen atmosphere. Polyimides containing heteroaromatic units, noncoplanar or rigid aromatic units show high heat resistances and high glass transition temperatures. However, polyimides that contain flexible linkages, such as ether units, show lower glass transition temperatures because of their relatively flexible polymer backbones. Pyridine rings increase the symmetry and aromaticity of the polymer and increase the thermal and chemical stability. In addition, pyridine rings help the polymer retain its mechanical properties at elevated temperatures.

Polyimides generally exhibit brown coloration due to charge transfer (CT) between the diamine donor moieties and dianhydride acceptor moieties. The optical and electrical properties of polyimides can be tailored by incorporating chromophores such as triphenylamine, carbazole, perylene groups, etc. Carbazole is a conjugated unit that has desirable photoconductivity and photorefractivity properties. In the field of electroluminescence, carbazole derivatives are often used as materials for hole-transport and in light-emitting layers because of their high charge mobility and thermal stability. Carbazole-containing polymers exhibit blue electroluminescence. Because the nitrogen atom improves the planarity of the biphenyl unit, the carbazole unit has a large band gap, and this leads to the observed electroluminescence.

The solubility of polyimides depends strongly on the chemical structure of the polymer. The two key factors in designing soluble and processable polyimides are ①reducing the rigidity or regularity of the backbone, and ②minimizing the density of imide rings along the backbone. In addition, aliphatic side chains have been incorporated into diamines to reduce the interaction between polyimide chains and increase the solubility. A large number of structural modifications have been attempted in previous decades, including incorporating thermally stable, flexible, or nonsymmetrical linkages in the backbone and introducing bulky substituents. Polyimides containing bulky, propeller-shaped triphenylamine units along the polymer backbone are amorphous and exhibit good solubility in many aprotic solvents, excellent film-forming capabilities, and high thermal stabilities. The incorporation of pendant cardo groups into the backbone of polyimides improves their solubility, processability and thermal stability. Furthermore, the tert-butylcyclohexylidene group, which can be considered an alicyclic cardo group, has been incorporated into polymer backbones to improve the polymer's processability.

Aromatic polyimides have excellent thermal stabilities and good mechanical properties, and they have been widely used in photoresists, liquid crystal alignments, gas separation membranes, composites, Langmuir—Blodgett (LB) films, blending applications, vapor phase depositions, electroluminescent devices, polyelectrolytes, fuel cells, electrochromic materials, nanomaterials

90

and polymer memory materials.

Photosensitive polyimides (PSPIs) are widely used in interconnects, multichip modules, protection layers, optical interconnects and resists because of their excellent thermal and chemical stabilities, low dissipation factors, and reasonably low dielectric constants.

The ultrathin mono-and multilayer films of polyimides have been successfully prepared using LB techniques in conjunction with a precursor method. Electrically insulating ultrathin LB films of polyimides (0.4 nm) have been successfully prepared on solid substrates.

The combination of carbon nanotubes and polyimides is expected to play an important role in the development of novel high-performance nanocomposites. There are two common processing techniques for fabricating the composites. One is to mix CNTs with the resin matrix in the melt state to form composites. The other technique involves dispersing the CNTs into a polymer solution, performing solution casting, and removing the solvent to obtain the composite.

Aromatic polyimides are well-known as high performance polymers due to their excellent thermal stability, mechanical and electronic properties; therefore, they are suitable to be used as a matrix resin in fiber reinforced composites. Usually, composites of carbon fibers and thermosetting polyimides are fabricated by routing an poly(amic acid) solution, because solubility of more than 30% is required to produce a prepreg. In this route, water generated as a by-product of imidization in the fabricating process might result in voids in the composites. Therefore, polyimides prepared through a polycondensation reaction require extremely severe processing conditions for molding. Polyimides can be used in novel applications in a variety of fields, and the excellent properties of polyimides will be developed and reported in the future.

3.9   Poly(ether ether ketone)

The polymer chemists were a little faster in attempting to develop useful polymers from the basic structure, and Carothers carried out some of the first work in the development of nylon in the 1930s. This basic research continued, and in the 1970s, Rose and his team at ICI began constructing polymers almost from the first reported principles. Their work was based on previously successful materials such as the polysulfones, and one of their first products was the aromatic polyether ketone family[15].

The preparation of these materials used a similar polyetherification process to the one that they had already developed for the polysulfones, and the structure of the resulting polymer was also very similar.

The linear aromatic polyether ketone family consists of several variations on the theme of repeated monomers of ether and ketone. The first one produced in the laboratory in 1977 was the polymer polyether ether ketone B (now simply known as PEEK) and can claim to be one of the first designer polymers.

The structure of PEEK is shown below:

$$\left[\begin{array}{c} \bigcirc - \overset{\displaystyle C}{\underset{\displaystyle O}{\parallel}} - \bigcirc - O - \bigcirc - O \end{array}\right]_n$$

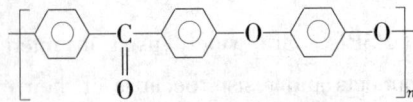

Polyether ether ketone (PEEK)

PEEK is available in a variety of grades for specific applications, and the main grades available are the following:

①Standard unfilled: This is a general-purpose grade, which has the highest elongation at break and also the lowest general mechanical properties (tensile strength, flexural strength and flexural modulus) at a given temperature. Unfilled PEEK also has the best impact properties of the PEEK range. Unfilled PEEK is compliant with FDA 21CFR 177.2145 for use in food contact applications.

②30% glass filled: The addition of glass fiber reinforcement greatly increases the general mechanical properties at a given temperature (tensile strength, flexural strength and flexural modulus), and reduces the elongation at break and impact strength at low temperatures. Glass fiber filled grades of PEEK also show reduced thermal expansion rates and are ideal for high temperature structural applications.

③30% carbon filled: The addition of 30% carbon fiber reinforcement further increases the general mechanical properties at a given temperature (tensile strength, flexural strength and flexural modulus), and further reduces the elongation at break and impact strength at low temperatures. Carbon fiber filled grades of PEEK also have much reduced thermal expansion rates and greatly improved thermal conductivity.

④Lubricated: The addition of 30% carbon/PTFE improves the tribological properties (friction and wear) of PEEK. The additives also improve the machinability and make this grade ideal for machined bearing parts.

Properties of PEEK is shown in Table 3.9. There are many superlatives that can be used to describe the properties of PEEK, and it is regarded by many as the best performing thermoplastic. For heat resistance PEEK has the reputation of providing the ultimate performance in a commercially available melt-processable thermoplastic.

Table 3.9 Properties of PEEK

Property	Approximate Value		
	Unfilled	30% Glass Fiber	30% Carbon Fiber
Tensile strength (at 23 ℃)	100 MPa	150 MPa	215 MPa
Tensile Modulus (at 1% strain at 23 ℃)	3.5 GPa	11.4 GPa	22.3 GPa
Elongation at Break (at 23 ℃)	34%	2%	1.8%
Flexural Strength (at 23 ℃)	163 MPa	212 MPa	298 MPa
Notched Impact Strength (at 23 ℃)	7.5 kJ/m^2	10.3 kJ/m^2	5.4 kJ/m^2

continued

Property	Approximate Value		
	Unfilled	30% Glass Fiber	30% Carbon Fiber
Specific Heat (Melt)	2.16 kJ/kg℃	1.7 kJ/kg℃	1.8 kJ/kg℃
Glass Transition Temperature	143 ℃	143 ℃	143 ℃
Heat Deflection Temperature	152 ℃	315 ℃	315 ℃
Coefficient of Thermal Expansion	$< T_g$ 4.7×10^{-5}/℃ $>T_g$ 10.8×10^{-5}/℃	$<T_g$ 2.2×10^{-5}/℃	$<T_g$ 1.5× 10^{-5}/℃
Long Term Service Temperature (Electrical)	260 ℃	240 ℃	N/A
Long Term Service Temperature (Mechanical-no impact)	240 ℃	240 ℃	240 ℃
Long Term Service Temperature (Mechanical-impact)	180 ℃	220 ℃	200 ℃
Specific Gravity	1.30	1.51	1.40
Water Absorption	0.50% (50% rh)	0.11% (50% rh)	0.06% (50% rh)
Transparency	Opaque (grey/brown)	Opaque (brown)	Opaque (black)

PEEK is a semi-crystalline material (typically 35%) and many of the properties derive from this degree of crystallinity in the final product. Degree of crystallinity is affected by the processing parameters, and therefore the properties quoted here should be regarded as being indicative only.

PEEK has greater strength and rigidity than many of the other engineering thermoplastics, making it tough over a wide range of temperatures.

It has good mechanical properties, including impact resistance, low wear rate, and a low coefficient of friction, but more importantly, these properties are also retained over a wide temperature range.

The thermal oxidative stability of PEEK is excellent and the material has an UL-rated continuous operating temperature of around 250 ℃ (depending on the grade used). PEEK has excellent resistance to burning and very low flame spread being rated as UL 94 V-0 for thicknesses down to 2 mm. The LOI (Limiting Oxygen Index) is 35% (depending on grade) and even when burning the material has one of the lowest smoke generation characteristics of the engineering thermoplastics.

PEEK has good dielectric properties, with high volume and surface resistivities and good dielectric strength. These properties are retained at temperatures as high as 200 ℃.

PEEK has outstanding chemical resistance and is extremely resistant to most organic and inorganic chemicals. It is dissolved or decomposed only by concentrated anhydrous or strong oxidizing agents.

The material has exceptionally good resistance to hydrolysis in hot water and remains unaffected after several thousand hours at more than 250 ℃ in pressurized water.

PEEK is not greatly resistant to UV radiation but has good resistance to beta and X-rays, as well as exceptional resistance to gamma rays (more than 1 000 Mrad without significant loss in mechanical properties). These properties allow for ease of sterilization, and coupled with good biocompatibility (USP Class VI), PEEK makes a strong candidate for medical applications.

Table 3.10 lists some of advantages and limitations of PEEK.

Table 3.10 **Advantages and Limitations of PEEK**

Advantages	Limitations
Excellent high temperature performance for all mechanical properties	Extremely high cost but the properties can justify this when it becomes almost the only polymer capable of being used
Excellent electrical performance at high temperatures	Limited supplier base
Excellent wear and abrasion resistance at high temperatures	Limited range of colors
Excellent chemical resistance at high temperatures	
Excellent gamma radiation resistance	

Despite the exceptional properties of PEEK, it can be processed by many of the traditional plastics processing methods and is manufactured in formats to suit these conventional processes. For injection molding, it is sometimes possible to use existing tooling if the tooling has suitable cavities and cores that will withstand the higher processing temperatures. Processing of PEEK is shown in Table 3.11.

Table 3.11 **Processing method of PEEK**

Processing Method	Applicable
Injection Molding	Yes
Extrusion (profiles, sheet and monofilament)	Yes
Compression Molding	Yes
Powder Coating	Yes

The outstanding mechanical properties of PEEK at high temperatures make it suitable for the most demanding applications, but the high cost sometimes limits applications to those where the properties are very necessary. Typical applications are the following:

①Automotive: Piston components and bearing linings. These applications are particularly suitable for the carbon fiber reinforced grades where the improved thermal conductivity means that heat is dissipated very quickly.

②Electrical engineering: Wire insulation for extremely high temperature applications, cable couplings, and connectors. In some of these applications, PEEK is even better than PTFE or other fluoropolymers because it has a greater resistance to cut-through by sharp edges.

③Appliances: Handles and cooking equipment.

④Medicine: Prosthetics, instruments, and diagnostics.

⑤Others: Aircraft parts and wire insulation, pump casings and impellers, monofilament for production of woven products for filters, belting, and meshes.

3.10   Fluoroplastics

Fluoroplastics are fluorocarbon-based polymers with multiple strong carbon-fluorine bonds. They represent a family of high-performance "super plastics" characterized by great strength, versatility, durability, outstanding electrical properties, and an unusually high resistance to chemicals and heat, with low smoke and flammability. PTFE is the best known member of this family, and one of the smoothest, toughest materials around.

Fluoropolymers are probably most recognized for their use as a coating for non-stick cookware or for insulating wire and cable, and have hundreds of other major applications that are essential to modern life. They are integral to a clean environment, to a globalized and connected world, and to the safety and security of families, workers and the public.

Fluoroplastics provide life-saving capabilities and reduce the risk of fire when used to insulate wire and cable placed in office ceilings. They protect workers from corrosive, alkaline, or acidic chemicals when used as a coating on protective garments, and are essential in the manufacture of durable fire-retardant fabrics. They protect the environment when used in solar collectors, anti-graffiti surface covers, and pollution control and alternative power applications. Fluoroplastics keep the world connected with qualities useful for data transmission cables, while also increasing electrical safety in cable connectors, cable jacketing, circuit breakers, heat-trace cable, stand-off insulators and tubing. The unique properties of fluoroplastics enable them to provide many different solutions in an increasingly complex world.

Among the fluoroplastics that contain monomers other than TFE, PCTFE was the first to be commercialized. Extensive development work was carried out on this polymer during World War Ⅱ in conjunction with the Manhattan Project. Compared to TFE resins, PCTFE is harder, more resistant to creep, and less permeable. It has the lowest permeability to moisture of any plastic.

Chlorotrifluoroethylene copolymerized with VDF provides an improved resin for film manufacture. However, ECTFE is the most important CTFE copolymer. These resins are similar in properties and uses to ETFE copolymers[16].

Poly(vinyl fluoride) (PVF) contains the smallest amount of fluorine (41.3 %) of any commercial fluoropolymer, but possesses many of the properties of more highly fluorinated polymers. It was commercialized in 1961 by DuPont under the trade name Tedlar. It is used mainly in coatings for metal, plastic, paper, and similar substrates to provide resistance to weather, chemicals, staining, and abrasion.

Poly(vinylidene fluoride) (PVDF) shares many of the characteristics of other fluoropolymers such as thermal and oxidative stability, as well as outstanding weatherability. However, the arrangement of alternate fluorine and hydrogen atoms leads to unusual polarity within the polymer chains, with a dramatic effect on dielectric properties and solubility. An unusual product made from PVDF is a film with piezoelectric properties. Copolymerization of VDF with a small amount (<15%) of HFP reduces stiffness and improves processability for certain wire coating applications.

Poly(tetrafluoroethylene) (PTFE) is a straight-chain polymer of tetrafluoroethylene (TFE) of the general formula:

$$\left[\begin{array}{cc} F & F \\ C{-}C \\ F & F \end{array}\right]_n$$

It has a high crystalline melting point (327 ℃), very high melt viscosity (10~100 GPa · s at 380 ℃), and a high maximum use temperature (>260 ℃). In addition, it exhibits unusual toughness down to very low temperatures (<−200 ℃); its molecular mass is extremely high (10^6~10^7). It is insoluble in all known solvents and resists attack by most chemicals. Dielectric loss is low, whereas dielectric strength is high; antistick and antikiction properties are most unusual. Although these properties give PTFE great commercial value, they also rule out processing by conventional thermoplastic techniques.

Manufacturers of PTFE resins include Ausimont (Algoflon and Halon), Daikin Kogyo (Polyflon), DuPont (Teflon), Hoechst (Hostaflon), ICI (Fluon), and Solvay Solexis (Fluoroplast), etc.

DuPont produces FEP resins under the Teflon trade name[17]. A similar product is manufactured by Daikin Kogyo of Japan and sold under the Neoflon trade name. The FEP resins are available in the form of extruded pellets (molding powders) and as an aqueous dispersion.

DuPont™ fluoroplastics are processed by various means, all of which depend on the intended application, end-use or properties desired. All standard operations—turning, facing, boring, drilling, threading, tapping, reaming, grinding, etc.—are applicable to PTFE resins. Special machinery is not necessary.

When machining parts from PTFE resins, either manually or automatically, the basic rule to remember is that these resins possess physical properties unlike those of any other commonly machined material. They are soft, yet springy. They are waxy, yet tough. They have the cutting "feel" of brass, yet the tool-wear effect of stainless steel. Nevertheless, any trained machinist can readily shape PTFE to tolerances of 0.001 in and, with special care, to 0.000 5 in.

Properties of fluoropolymers that have led to applications include chemical resistance, thermal stability, cryogenic properties, low coefficient of friction, low surface energy, low dielectric constant, high volume and surface resistivity, and flame resistance. Fluoropolymers are used as liners (process surface) because of their resistance to chemical attack. They provide durable, low maintenance, and economical alternatives to exotic metals for use at high temperatures without introducing impurities. Electrical properties make fluoropolymers highly valuable in electronic and electrical applications as insulators, e.g., FEP in data communications[6].

In automotive and office equipment, mechanical properties of fluoropolymers are beneficial in low-friction bearings and seals that resist attack by hydrocarbons and other fluids. In food processing, the Food and Drug Administration approved fluoropolymer grades are fabrication material for equipment due to their resistance to oil and cleaning materials, and their anti-stick and low friction properties. In houseware, fluoropolymers are applied as nonstick coatings for cookware and appliance surfaces. Medical articles such as surgical patches and cardiovascular grafts rely on the long-term stability of fluoropolymers as well as their low surface energy and chemical resistance.

For airports, stadiums, and other structures, glass fiber fabric coated with PTFE is fabricated into roofing and enclosures. PTFE provides excellent resistance to weathering, including exposure to UV rays in sunlight, flame resistance for safety, and low surface energy for soil resistance, and easy cleaning.

Some applications of PTFE is shown in Figure 3.14.

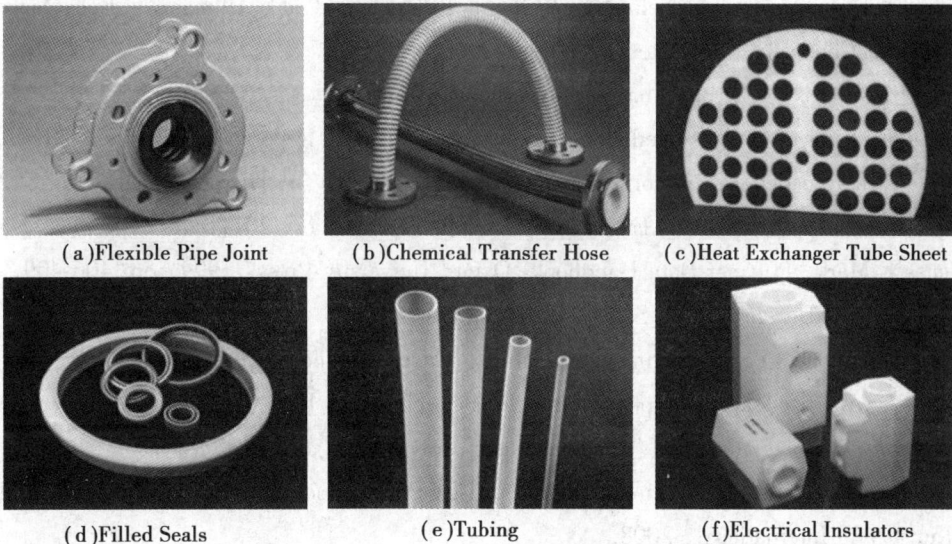

| (a)Flexible Pipe Joint | (b)Chemical Transfer Hose | (c)Heat Exchanger Tube Sheet |
| (d)Filled Seals | (e)Tubing | (f)Electrical Insulators |

Figure 3.14   Typical applications of PTFE

Collection of Exercises

1. Write down the chemical structure of PET?

2. What are the structure characteristics of PTFE? Why it has good auto-lubricating property?

3. What are the main characteristics of Nylon families?

4. Compare the properties of Nylon with POM, and find out their potential application fields.

5. Why does PPS show an excellent heat resistance?

6. List the main applications of PPS.

7. Why must polyacetal (POM) be capped?

8. What is the advantage of a clear polycarbonate sheet over a sheet of poly (methyl methacrylate)? Why?

9. What polymer is used in prototype automobile combustion engines? Why?

10. Which has the higher melting point: nylon 66 or Kevlar?

REFERENCES

[1] Charles E. Carraher Jr., Giant Molecules. Essential Materials for Everyday Living and Problem Solving, Wiley-Interscience, 2003.

[2] Horst Köpnick, Manfred Schmidt, Wilhelm Brügging, et al., Ullmann's Encyclopedia of Industrial Chemistry (Polyesters), Wiley, 2000.

[3] http://www.wisegeek.com/what-is-polycarbonate-resin.htm, wiseGEEK 网站资料.

[4] James D. Virosco, Polyoxymethylene (Polyacetal) 2011S1 Report, Nexant Inc., 2012.

[5] http://www.boedeker.com/noryl_p.htm, Boedeker Plastics Inc.网站资料.

[6] Myer Kutz, Applied Plastics Engineering Handbook, Elsevier, 2011.

[7] James E.Mark, Polymer Data Handbook. Oxford University Press, 1999, pp. 406-409.

[8] Udel ® PSU Design Guide, Solvay 公司技术资料.

[9] Polyethersulfone (PES) Technical Literature, Mitsui Chemicals, Inc., 2011.

[10] http://www.boedeker.com/pes_p.htm, Boedeker Plastics Inc.网站资料.

[11] http://www.solvay.com/EN/Homepage.aspx, Solvay 公司产品技术资料.

[12] Dipl.-Ing. Nicolas Inchaurrondo-Nehm, High-Performance plastics (Polyarylsulfones), Kunststoffe International, 2008.

[13] Varun Ratta, Crystallization, Morphology, Thermal Stability and Adhesive Properties of Novel High Performance Semicrystalline Polyimides, Dissertation, Virginia Polytechnic Institute

and State University, 1999.

[14] Der-Jang Liaw, Kung-Li Wang, Ying-Chi Huang, et al. Advanced polyimide materials: syntheses, physical properties and applications, Progress in Polymer Science, 37: 907-974, 2012.

[15] http://www.zeusinc.com/pdf/Zeus_PEEK.pdf, Zeus Industrial Products, Inc., 2005.

[16] Klaus Hintzer, Tilman Zipplies, D. Peter Carlson, Ullmann's Encyclopedia of Industrial Chemistry (Fluoropolymers), Wiley, 2014.

[17] http://www.rjchase.com/ptfe_handbook.pdf, Dupont 公司产品技术资料.

CHAPTER 4
THERMOSETS

4.1　Introduction

According to plasticity, plastics can be divided into thermoplastics and thermosetting plastics. Generally, thermoplastics can be recycled while thermosetting plastics cannot. The characteristics of the thermosetting plastics is chemical reactions and hardening occurs when under a certain temperature and after heating, pressing or adding hardener. After hardening the chemical structure of plastic changes, and the plastic becomes hard, hardly dissolves in solvents, no longer soften after heating, but decomposes if the temperature is too high. Thermoplastic resin molecular chains are linear or branched chain structure without chemical bonds between molecular chains, they soften and flow after heating. The process of hardening as it cools is physical change.

The commonly used thermosetting plastics are phenolic resin, urea-formaldehyde resin, melamine resin, unsaturated polyester resin, epoxy resin, silicone resin, polyurethane, and so on.

4.2　Processing methods for thermosets

Thermosetting molding compounds (thermosets) that undergo a chemical reaction or cure (called polymerization or molecular growth) during the molding operation, include phenolic (phenol-formaldehyde), urea, melamine, melamine-phenolic, diallyl phthalate, alkyd, polyester, epoxy, and silicones. Thermosetting molding compounds processed from the individual heat-reactive resin systems are available in a wide range of formulations to satisfy specific end-use requirements. Depending on the type of material, products may be supplied in granular, nodular, flaked, diced, or pelletized form. Polyester materials are supplied in granular, bulk, log, rope, or sheet form, and polyurethanes are

100

made in many forms, ranging from flexible and rigid foams to rigid solids and abrasion-resistant coatings [1].

Thermosets, when placed in a heated mold under pressure, will conform to the shape of the mold and cure into a hard infusible product. Successful plastics molding is dependent on good mold design and construction, the mold temperature, material temperature, molding pressure, etc.

The oldest and simplest method of processing thermoset molding materials is Compression Molding (Figure 4.1). The mold consists of a cavity side, with one or more cavities and a force side (Figure 4.2). The mold is heated by either electric cartridge heaters, steam or oil to a temperature range of 165 ~ 182 ℃ for phenolic molding compounds, 150 ~ 177 ℃ for melamine-phenolic molding compounds,

Figure 4.1   Compressing mold machine

163 ~ 182 ℃ for granular polyester molding compounds or 143 ~ 171 ℃ for BMC polyester molding compounds [2].

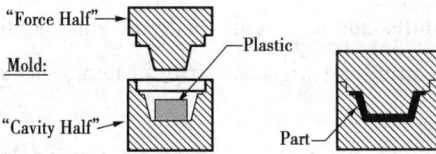

Figure 4.2   Compression molding

All of the thermoset compounds, except epoxies and silicones, may be molded by the following methods: compression, transfer, thermoset injection, and the runnerless injection/compression process. This chapter will only discuss compression and transfer molding. Table 4.1 lists factors to be considered in the selection of compression or transfer molding.

The design and construction of the mold is the single most important factor in a plastics molding project. Without a mold built and engineered to produce good molded parts in an economical fashion, other factors are of little importance.

Table 4.1   Selection of molding method-compression or transfer

Factors to Conside: Advantages—Limitations	Compression	Transfer
Close tolerances, projected area	●	
Close tolerances, over flash line, minimum flash		●
Lowest mold shrinkage	●	
Uniform shrinkage, all directions	●	
Maximum uniform density	●	
Reduced cure, thick sections		●
No weld lines, less mold-in strains	●	
Small holes, longer length, through holes		●

continued

Factors to Consider: Advantages—Limitations	Compression	Transfer
Extremely thin mold sections, telescoping		•
No venting problems	•	
Impact strength	•	
Molds with movable sections or cores		•
Molded-in inserts		•
Large projected area parts	•	
Lowest mold-flash scrap	•	
Generally less mold maintenance		•
Gate or sprue removal necessary		•
Maximum number cavities per clamp force	•	
Molds erosion, sprues, runners, gates		•
Generally higher mold cost		•

Thermoset molding compounds may be molded in a temperature range of $141 \sim 204$ °C. Material suppliers should be consulted for recommended temperatures for a specific material and molding method. Molds may be heated by steam, hot oil, electric cartridge or strip heaters, or any combination of these.

In compression molding, the thermoset compound is placed in the open heated mold. The material may be in powder form or as a preform, a cold pressed slug that contains the exact charge weight required. As the mold closes, the heat and pressure cause the material to flow, compressing it to the required shape and density as defined by the mold. Continued heat and pressure produce the chemical reaction (polymerization or cure) that hardens the material. The thinner the part, the shorter the cure; conversely, thicker pieces take longer to cure. Part design should have as uniform a wall thickness as possible.

The mold is fastened in a vertical molding press, either up-acting or down-acting, usually hydraulically driven. Small bench presses may be air driven. Presses may be self-contained or on a common hydraulic system. They have provision for an ejection system for the parts, usually both up-acting and down-acting. The press operation may be either automatic or semi-automatic, in sizes up to 2 000 tons.

Preform presses, high-frequency preheaters, and preheat extruders are commonly used as auxiliary equipment in the compression-molding process.

The mold consists of two halves, one containing the cavity or cavities (the female section) and one containing the force or forces (the male section). Each is mounted on press supports or grids which are in turn fastened to the stationary or moving platens. Generally, the cavities are in the lower half to permit easy loading of the molding compound. This operation may be manual or automatic. In the case of automatic operation, movable loading trays are incorporated, in conjunction with trays or forks to receive molded parts from the mold. A predetermined amount of molding material is placed into the open mold. By closing the mold and compressing the material, the desired shape is achieved.

Both thermosets and thermoplastics may be compression molded; the process is used mostly for thermosets and thermoplastics containing significant amounts of fillers which reduce their viscosity (flowability), required for injection molding. (Examples: phenolics with up to 60% mineral fillers, for insulators; polysulfones up to 80% ceramic, for engineering uses.) Flashing is necessary to allow the air or gases to escape.

A typical cycle, with the mold at recommended temperature and with adequate pressure available, would proceed as follows:

①Air-clean the mold of all flash or foreign matter.

②Load the material into the cavities.

③Close the mold completely; or before closing it, interject a brief 'breathe cycle' by opening the mold slightly to release any air and gases trapped in the molding compound.

④Complete the cure time.

⑤Open the mold and activate the knock-out assembly.

⑥Remove the molded parts.

⑦Clean the mold with an air blast.

The cure duration is dependent on the type of molding compound, mold temperature, pressure on the material, and material temperature. Cross-sections $0.125 \sim 0.500$ in. ($3.18 \sim 12.7$ mm) thick may cure in 30 seconds $\sim$ 2 minutes when preheated material is used (which is always desirable).

For thermosets, the material can be cold, but more often it is preheated close to the "setting" temperature. The mold is heated using steam or electric heaters, and remains closed until the part is "cured" or "set"; the hot part is then ejected. Typical parts produced using this technique are tires, components for the electrical industry, dinnerware, and under-the-hood automotive parts.

For thermoplastics, the hot (melted) material is placed into the relatively cold mold; the mold closes and compresses the material into the desired shape. When the part has cooled down enough to be handled without deformation, it is ejected.

With the materials processed by compression molding, the mold is open while the material is loaded. During the closing, it is unavoidable to have excess material escape at the parting line (flash). It is necessary to provide an excess of material with every shot, which is wasted (scrap) with thermosets, but can be reused with thermoplastics. In either case, the flash must be removed to provide a finished part, either by tumbling manually (parts permitting) or mechanically.

Transfer molding is a method of molding specific parts using a mold with two halves that is closed before any material is introduced. The material is loaded into a pot or transfer sleeve, and transfer pressure is applied to cause material to flow into the closed section of the mold. In a single-cavity mold, the material flows generally through a sprue bushing and is gated directly into the part. In the case of a multi-cavity mold, it flows from a sprue bushing or transfer sleeve into a runner system and is gated into each cavity and part.

There are two distinct transfer methods of molding. One is known as pot-type transfer, and the other is the plunger transfer method.

Pot-type transfer molding is generally done in a bottom-clamp compression press. Plunger transfer molding is done in a vertical press. A transfer press has a hydraulic clamp cylinder with a separate transfer cylinder applying pressure in the direction opposite to the clamp pressure. For automatic operation, a top-clamp force and a bottom-transfer force are desirable for ease of loading preheated preforms. Plunger-transfer presses are generally self-contained and have provisions for top and bottom knockout systems that are available for semiautomatic or automatic operations. Users should contact press manufacturers or molders for information on available press sizes.

Depending on whether the molding method is pot-type or plunger, one of two mold types is employed in transfer molding. The mold cavity is closed (clamped), and connected at a gate via a runner to the transfer pot. The material is loaded into the pot and a plunger drives the material into the cavity. There is no flash on the part, but the runners and a certain amount of scrap (cull) remaining in the bottom of the pot must be removed before the next shot. This method is used almost exclusively for thermosets, requiring the pot to be heated. To reduce the molding time, the material charged into the pot should be well preheated, and as close as possible to the curing temperature of the material.

A pot-type transfer mold as shown in Figure 4.3 generally has one cavity. The mold consists of two halves with the cavity section assembled to the lower mounting plate which is fastened to the supports or grids and bolted to the movable lower platen. The lower platen is moved up and down by the clamp ram. The force section is assembled to the lower surface of the movable floating platen. The pot or chamber is contained in the upper area. These components are fastened together as a complete assembly. The plunger that enters the pot area is mounted to the head of the press or to the grids.

In a typical cycle using the recommended mold temperature, the operator loads preheated preforms or extrudates into the pot area. Then the press is activated upward using low pressure, and the press picks up the floating member (Figure 4.3(a)) which engages the plunger. High pressure is applied, forcing the material through the sprue bushing directly into the cavity and force area (or through a diaphragm gate if it is a circular part). The cure cycle is completed under pressure, the clamp ram is moved downward, and the mold opens. The pot and plunger separate, and the movable floating platen is pulled away from the lower half of the mold by rods fastened to the head of the press. The part-removal assembly raises the part from the cavity, and the operator removes the part. The operator removes cull and sprue from the plunger, uses an airblast to clean the remainder of the mold, and places the preforms in a preheater or activates the extruder for the following cycle, as shown in Figure 4.3(b).

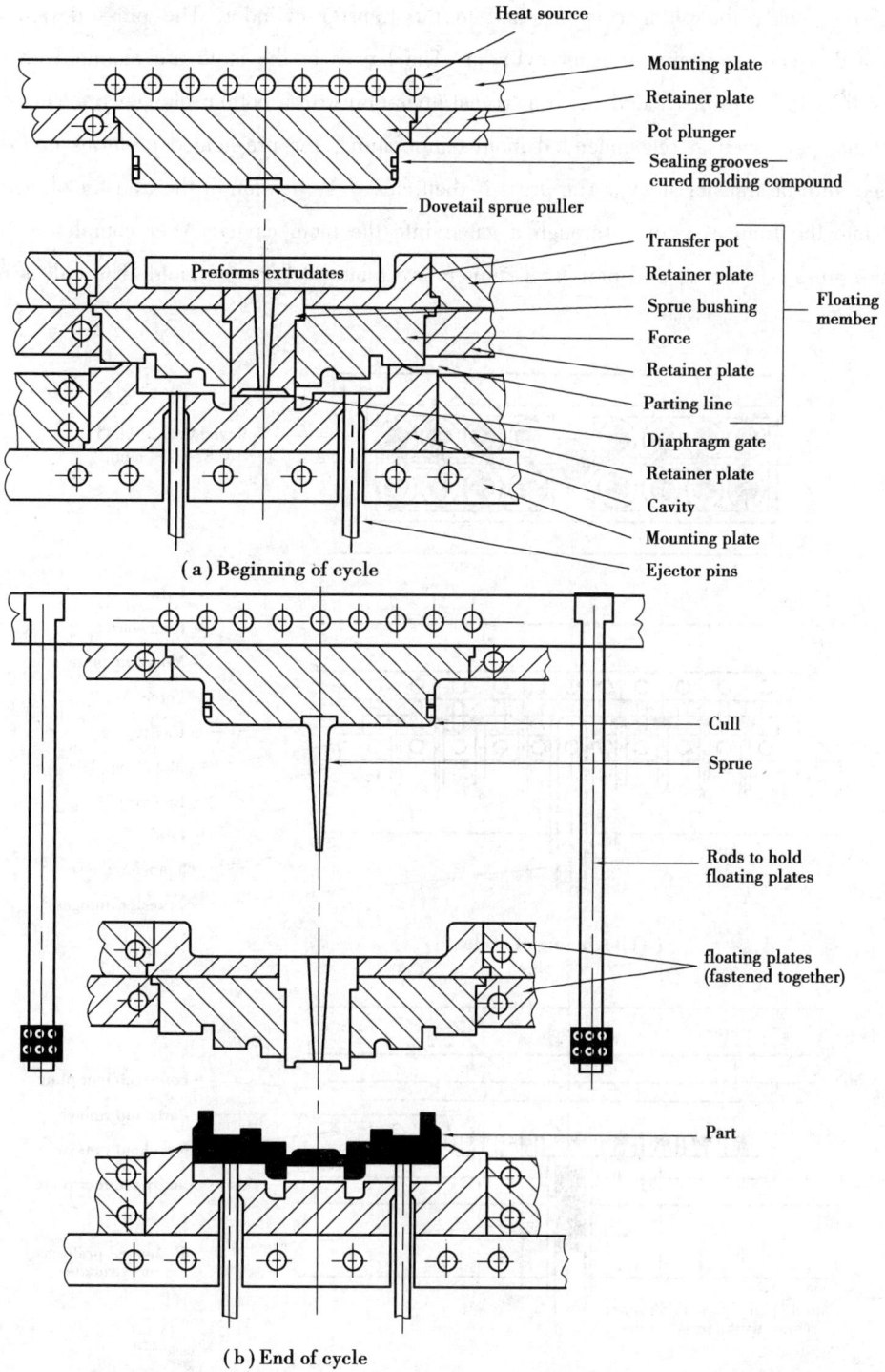

Heat source
Mounting plate
Retainer plate
Pot plunger
Sealing grooves—cured molding compound
Dovetail sprue puller

Preforms extrudates
Transfer pot
Retainer plate
Sprue bushing
Force
Retainer plate
Parting line
Floating member
Diaphragm gate
Retainer plate
Cavity
Mounting plate
Ejector pins

(a) Beginning of cycle

Cull
Sprue
Rods to hold floating plates
floating plates (fastened together)
Part

(b) End of cycle

Figure 4.3   Pot-type transfer mold(Courtesy R. W. Bainbridge)

A plunger transfer mold (Figure 4.4) consists of two halves, one containing the cavity or cavities and one containing the force or forces. The transfer sleeve and plunger are located in the center of the mold; the plunger is fastened to the transfer cylinder. The press design dictates location of the clamp ram and transfer cylinder. The halves of the mold are mounted on grids or support pillars in the proper location. In a typical operation with a bottom plunger transfer press and with molding performed at recommended mold temperature, the pre-heated preforms or extrudates are loaded into the transfer sleeve. The press is then closed. Activation of the transfer plunger forces material into the runner system, through a gate, into the mold cavity. After completing the cure cycle, the press is opened, and parts and runners are removed from the mold. The cull is removed

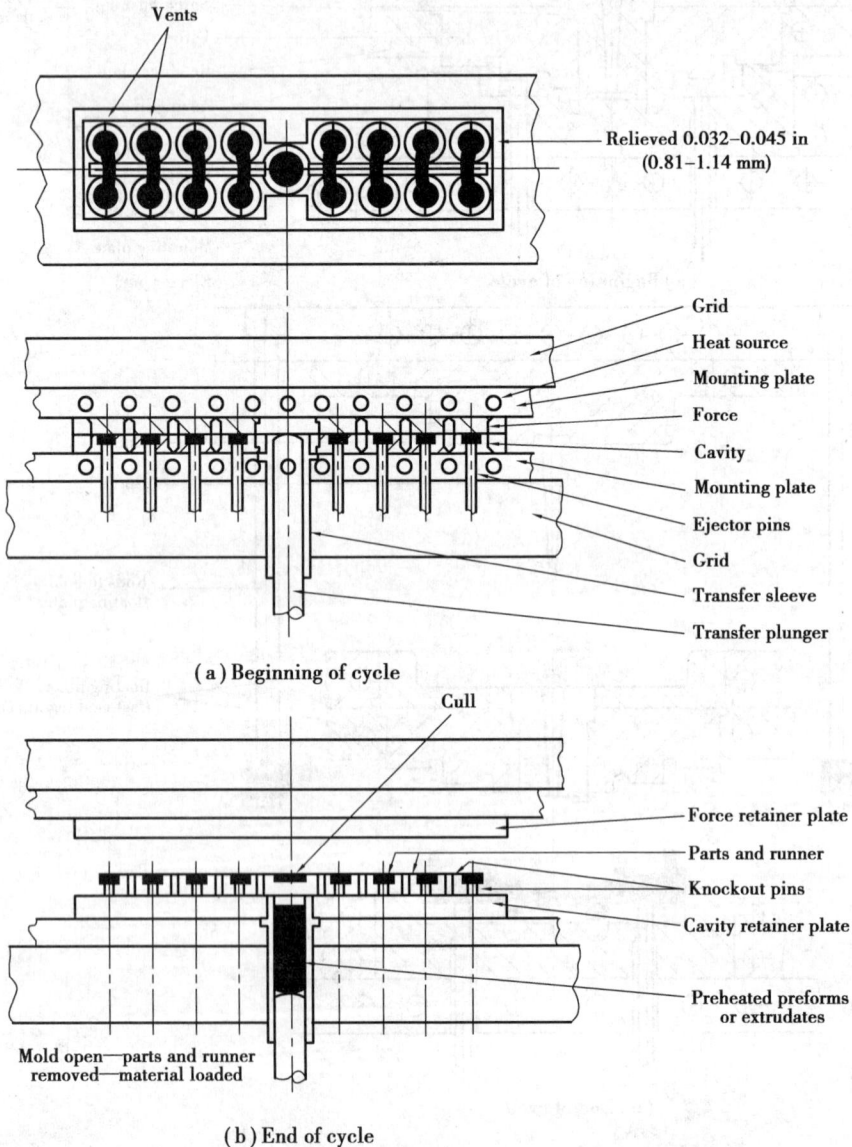

Figure 4.4   Plunger-type transfer mold(Courtesy R. W. Bainbridge)

from the top of the transfer plunger, and the transfer plunger is activated downward. An airblast is used to clean the mold and vents. The final step is to activate the preheater or extrudate equipment. Figure 4.4 shows a plunger transfer mold at the beginning (a) and the end of a cycle (b).

One of the key advantages of transfer molding over compression molding is that different inserts, such as metal prongs, semiconductor chips, dry composite fibers, ceramics, etc., can be placed/positioned in the mold cavity before the polymer is injected/drawn into the cavity. This ability makes transform molding the leading manufacturing process for integrated circuit packaging and electronic components with molded terminals, pins, studs, connectors, and so on.

In the composite industry, fiber-reinforced composites are often manufactured by a processed called Resin Transform Molding (RTM). Layers of textile preforms (long fibers woven or knitted in patterns) are pre-arranged in the mold. The resin is then injected to impregnate the performs. Vacuum is often used to avoid air bubbles and help draw the resin into the cavity. In addition, the resin used has to be relatively low in viscosity.

Although transfer molding can also be used for thermoplastics, the majority of the materials used in this process are still thermosets.

4.3   Phenolic resins [3]

Phenol-formaldehyde (PF) resins are obtained in the reaction of phenol and formaldehyde in aqueous alkaline solutions. This type of resin has a wide range of commercial applications in industrial products such as molding compounds, coatings, flame retardants, and wood adhesives.

The first synthetic resins and plastics were produced by polycondensation of phenol with aldehydes. In 1909 Baekeland made the first plastics. He carried out the polycondensation of phenol and formaldehyde to form cross-linked thermosets over several steps. Besides the production of plastics, phenolic resins were sought as a replacement for natural resins, which were then used on a large scale for oil varnishes. Between 1928 and 1931 phenolic resins gained increased importance through the treatment of resols with fatty oils to give air-drying varnishes. The main problem, an inadequate compatibility of phenolic resins with other varnish raw materials, was solved by using alkylphenols or by etherification of the hydroxymethyl groups of resols with monohydric alcohols.

These varnish applications and the use of phenolic resins as thermosets and electrical insulating materials were the main application areas. However, other polycondensates and, above all, polymers increasingly limited the market for phenolic resins from the mid 1930s onwards. The industrial development of phenolic resins is still continuing despite the long history. Their importance is likely to remain considerable because the raw materials can be obtained at reasonable cost from both petroleum and coal. Phenolic resins can be used as raw materials for synthetic fibers and in photoresists for the production of microchips which characterizes the continuing relevance of this group of resins.

Phenolic resins are classified as novolacs and resols. In resols the polycondensation is base-catalyzed and has been stopped deliberately before completion. Characteristic functional groups of this class of resins are the hydroxymethyl group and the dimethylene ether bridge. Both are reactive groups. During processing the polycondensation can be restarted by heating and/or addition of catalysts i.e., resols are self-cross-linking. In the case of novolacs the polycondensation is brought to completion. The molecular growth of these thermoplastic synthetic resins is limited by addition of a substoichiometric amount of the aldehyde component. Novolacs are phenols that are linked by alkylidene (usually methine) bridges, without functional groups (apart from the phenolic hydroxyl groups), and cannot cure on their own. However, novolacs can be cross-linked by addition of curing agents, such as formaldehyde or hexamethylene-tetramine, and give end products similar to resols.

The classification of phenolic resins into novolacs and resols is only strictly valid if phenols which are trifunctional towards formaldehyde are used as starting material, because resols from bifunctional phenols cannot cross-link by themselves. Nevertheless, the polycondensates from substituted phenols are differentiated according to their characteristic groups as alkylphenol novolacs (alkylidene bridge) or alkylphenol resols (hydroxymethyl group, dimethylene ether bridge).

The third large group are phenolic resins modified by natural resins. Besides phenolic hydroxyl groups, they contain double bonds, ester links, and carbonyl groups. Novolacs. The first step in phenolic resin polycondensation is always the electrophilic attack of a carbonyl compound (generally formaldehyde) on the para-and/or ortho-positions of a phenol molecule (acid catalysis, Equation 4.1) or a phenolate anion (base catalysis, Equation 4.2).

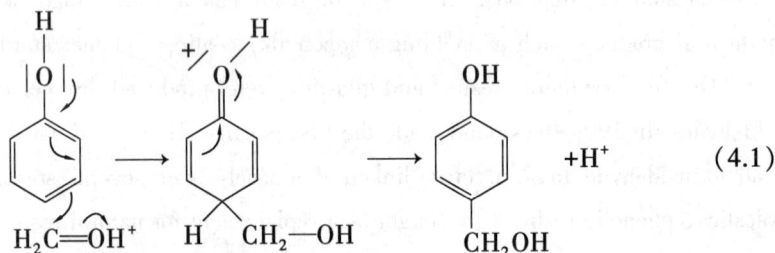

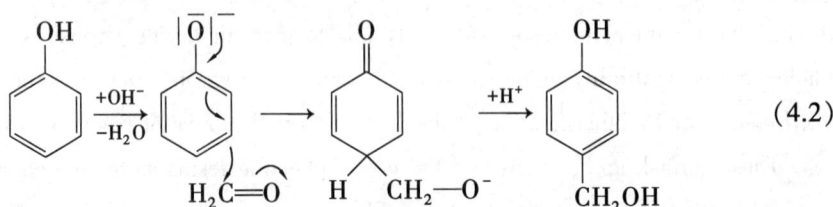

$$\text{(4.1)}$$

$$\text{(4.2)}$$

Since hydroxymethyl-substituted phenols are more reactive than phenol itself, the hydroxymethylation continues. The hydroxymethyl compounds formed are unstable in acidic medium and are rapidly converted into compounds linked by methylene bridges. This reaction also occurs in both the ortho-and para-positions. In basic media hydroxymethyl groups can be stable. At higher temperatures, however, they react with the formation of methylene bridges according to Equation

108

4.3. To hinder polyalkylation by cross-linking which would make further processing more difficult or impossible, less than one mole of formaldehyde must be added per mole of phenol.

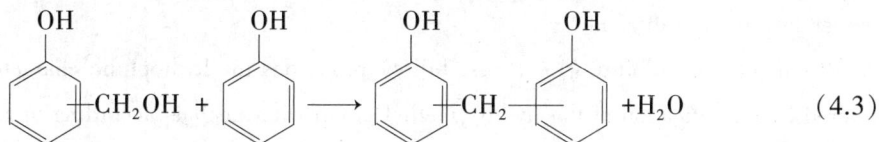

$$\text{(structure: phenol-CH}_2\text{OH} + \text{phenol} \longrightarrow \text{phenol-CH}_2\text{-phenol} + H_2O) \qquad (4.3)$$

Novolacs are sometimes used as chemically unmodified synthetic resins. Their main application is based, however, on their capability to undergo cross-linking with hexamethylenetetramine. The reaction occurs at ca. 150 ℃ according to Equation 4.4.

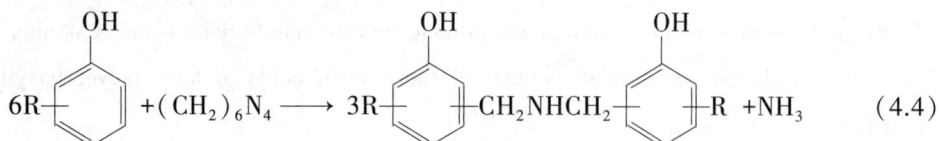

$$6R\text{-phenol} + (CH_2)_6N_4 \longrightarrow 3R\text{-phenol-}CH_2NHCH_2\text{-phenol-}R + NH_3 \qquad (4.4)$$

In a strongly acidic medium, hydroxymethyl groups are rapidly converted into methylene bridges. Therefore, the synthesis of resols can only be catalyzed by bases or salts of weak acids or bases. In analogy to the novolacs, the hydroxymethyl groups are formed in the ortho-or para-positions (Equation 4.2). At temperatures above ca. 40 ℃ the hydroxymethyl groups can react to form dimethylene ether bridges with elimination of water, according to Equation 4.5. No catalyst is needed for this reaction. The dimethylene ether bridges formed are more stable in the ortho-than in the para-position. They can be converted into methylene bridges with elimination of formaldehyde (Equation 4.6). The formaldehyde liberated is then available for the formation of new hydroxymethyl groups, if a suitable catalyst is present.

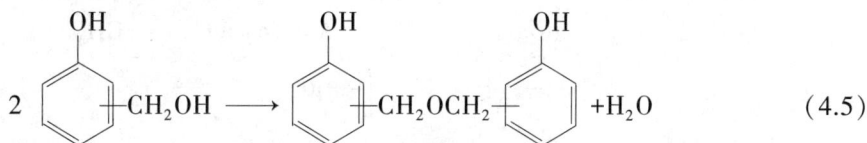

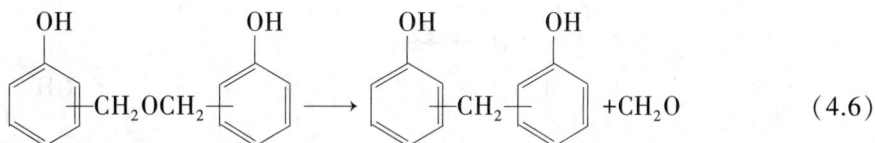

$$2\ \text{phenol-}CH_2OH \longrightarrow \text{phenol-}CH_2OCH_2\text{-phenol} + H_2O \qquad (4.5)$$

$$\text{phenol-}CH_2OCH_2\text{-phenol} \longrightarrow \text{phenol-}CH_2\text{-phenol} + CH_2O \qquad (4.6)$$

The hydroxymethyl groups of resols can also condense directly with other phenol molecules according to Equation 4.3. In resols therefore, three different types of formaldehyde-derived moieties occur: ①relatively stable methylene bridges (the only type of linkage in novolacs), ②hydroxymethyl groups, which are both capable of condensation reactions, and ③dimethylene ether bridges.

The structure of resols depends not only on the choice of raw materials and their molar ratios, but also on the temperature of formation, concentration of raw materials, presence or absence of solvents, type of catalyst, and catalyst concentration. These parameters determine the structure to a much greater extent than in the case of novolacs.

The catalyst-containing synthetic resins formed are ready to be used industrially. However, the catalyst can also be removed, either before application or during production, because no catalyst is required for the conversion of hydroxymethyl groups into dimethylene ether bridges, a rise in temperature being sufficient.

To affect compatibility (i.e., miscibility) properties the hydrophilic character of resols can be lowered by etherification of the hydroxymethyl group with alcohols according to Equation 4.7.

$$\text{[OH-phenyl]}\text{—CH}_2\text{OH} + \text{ROH} \longrightarrow \text{[OH-phenyl]}\text{—CH}_2\text{OR} + \text{H}_2\text{O} \tag{4.7}$$

Rosin, a natural resin, contains abietic acid and its double bond isomers as main components. Resols react with the unsaturated centers of these resin acids to form polycarboxylic acids with methylene bridges.

These condensation products from rosin and phenol-formaldehyde resin are known as albertol acids. They can be converted by esterification with polyols, or by salt formation, into higher molecular mass products which are readily soluble in nonpolar solvents but can release the solvent rapidly.

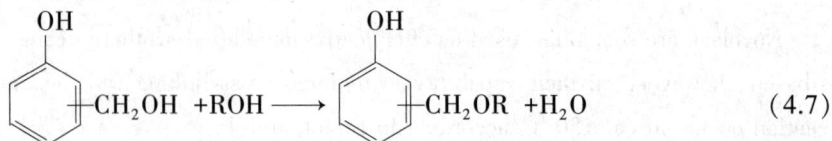

Abietic acid

$$\downarrow -2\text{H}_2\text{O}$$

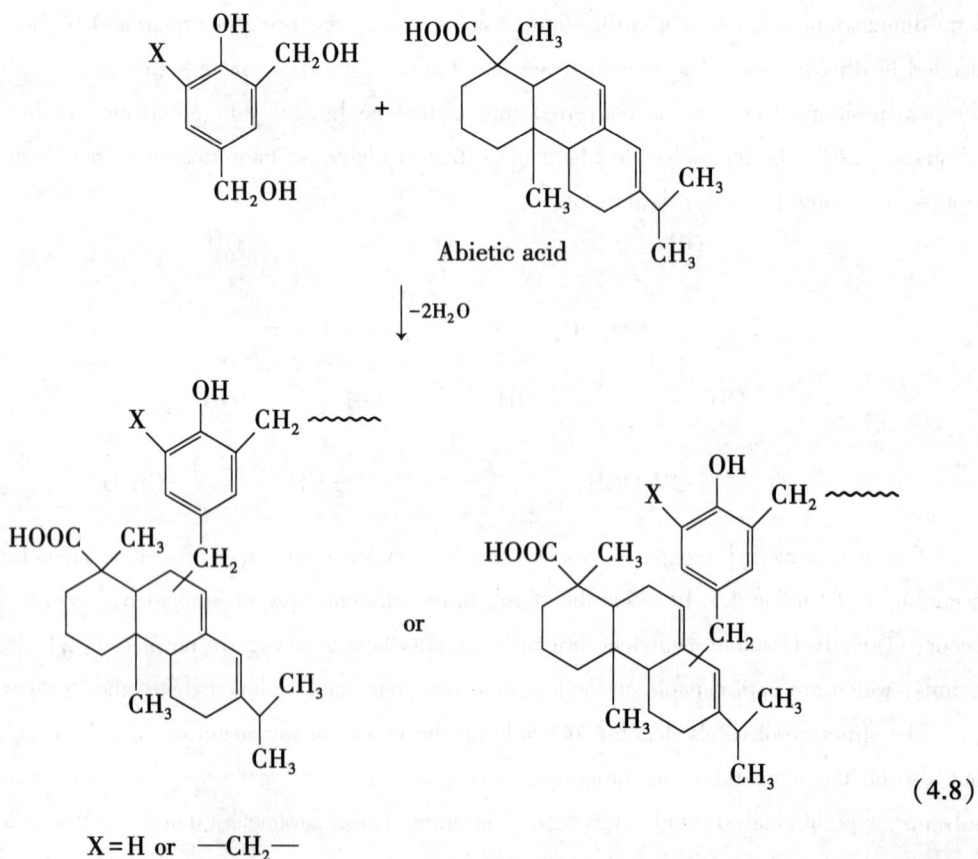

or

$$\tag{4.8}$$

X = H or —CH₂—

Albertol acids are also obtained by direct condensation of rosin, phenols, and formaldehyde.

Resols also undergo analogous reactions with other natural or synthetic unsaturated compounds such as fatty oils, rubbers, and polymer oils. A limited increase in the molecular mass or cross-linking can thus be achieved. Whether a particular reaction can be carried out successfully depends on the ratio of the rates of auto-condensation of the starting materials to co-condensation with the other reaction partners, but particularly on their mutual compatibility. Phenolic resins are therefore often classified as "water-soluble", "alcohol-soluble", "oil-soluble" etc.

Compatibility of resols with other components can be produced in many ways, e.g., ① by using ring-alkylated phenols as the raw materials; ② by etherification of the hydroxymethyl groups with alcohols; ③ by co-condensation of the resol with natural resins; or ④ by a combination of these measures.

Phenolic resins are yellow to brown in color, and the coloration can be very intense. Pale phenolic resins become colored immediately after production during storage or processing. The coloration is less intense only in the case of phenolic resins from para-alkyl-substituted phenols.

Phenolic resins which are not cross-linked are commercially available as solids or solutions. For particular applications, e.g., in thermosets, the polycondensation can be driven so far that the resins are no longer soluble but can only be swelled by organic solvents. The softening point of solid resins can be determined by the capillary melting point according to DIN 53 244, by the ring and ball method, or similar procedures. These temperatures are not melting points in the thermodynamic sense. They characterize a lowering of viscosity caused by a rise in temperature, as a result of which previously crushed resin particles can be observed to coalesce or another change in form occurs.

Cross-linked phenolic resins are hard substances which only have a small fracture strain and cannot be melted. Decomposition reactions begin at $120 \sim 250$ ℃, depending on the molecular structure. There are, however, also types of phenolic resin (phenolic ether resins) which are stable for some time up to 300 ℃.

Phenolic resins can be plasticized. Their compatibility with plasticizers can be adjusted by introduction of hydrophilic or hydrophobic groups.

For cross-linking, novolacs are processed together with curing agents, mainly hexamethy-lenetetramine. Cross-linking occurs at a sufficient rate at $140 \sim 160$ ℃ and can be carried out within a few minutes, particularly if a part of the polycondensation reaction has already taken place.

Novolacs are sometimes cross-linked with resols. This reaction gives resites with high hardness, high stability, but a very low fracture strain.

The deflection under load is tested according to Martens. According to DIN 53 458 and DIN 53 462, the fatigue strength at elevated temperatures is determined from properties such as the weight loss after storage in air or under an inert gas. Cross-linked phenolic resins are much less inflammable than thermoplastics. Phenolic resins can be rendered virtually nonflammable by the addition of usual plastics additives such as phosphates, borates, halo compounds, red phosphorus, phosphoric acid esters, or antimony trichloride. This is particularly important if other inflammable

111

substances are used as fillers or for reinforcement.

Phenolic resin molding materials are produced with heated rollers or extruders. Treatment of a mixture of novolac, hexamethylenetetramine, fillers, and additives leads to an intermediate, the actual molding material. The polycondensation is driven so far that the molding material in the final molding step can still flow, but so that the final cross-linking occurs very quickly. Phenolic resin thermosets are standardized according to DIN 7708, part 2. The application properties of the final products depend very much on the choice of filler material. The processing properties, in contrast, are affected primarily by the novolacs used. Especially important processing properties are good flow characteristics and rapid curing. Molding technology which predominated until the end of the 1960s has in the meantime been substituted by transfer molding and injection molding, particularly in the mass production of less complicated preforms. Thus the expression "thermosetting molding materials" has come into common use.

①Grinding Wheels: In the production of phenolic resin-bound grinding wheels, the grinding material (usually corundum of various granularities) is impregnated with a liquid phenol resol and mixed with a ground, pulverized mixture of phenol novolac and hexamethylenetetramine. The binder cures to a three-dimensional cross-linked resite under a carefully controlled temperature and pressure program. No defects may occur in this process because grinding wheels are exposed to high thermal and mechanical stresses (DIN 69 100, DIN 69 111).

②Friction linings: Brake linings and clutch facings are made of reinforced phenolic resinbound thermosets. Novolac-hexamethylenetetramine mixtures can also be used here as binders. For better heat conduction, copper wire or copper gauze is often incorporated. Novolacs modified with alkylphenols or cashew oil are used to adjust hardness and lubricating properties, particularly in brake linings.

③Reinforcing Resin for Rubber: The hardness of rubber is increased by the incorporation of novolac-hexamethylenetetramine mixtures. This hardness can be so great as to enable fabrication of solid, tough molded articles, for example for bodywork parts. Often, however, as in car tire mixtures, only slight improvement of the rubber hardness is aimed at. The reinforcing effect is thought to be caused by strong intermolecular interactions between the cured, duroplastic phenolic resin and the rubber-elastic vulcanized product. Both types of macromolecule are not bonded by covalent bonds.

4.4　Amino resins

Only about 10% of the total urea production is used for amino resins, which thus appear to have a secure source of low cost raw material. Urea is made by the reaction of carbon dioxide and ammonia at high temperature and pressure to yield a mixture of urea and ammonium carbamate; the latter is recycled [4].

$$2NH_3 + CO_2 \rightleftharpoons CO(NH_2)_2 + H_2O \tag{4.9}$$

Amino resins are the resins produced by the reaction between amino group-containing compounds and formaldehyde. The most popular amino resins are urea-formaldehyde (UF) resins and melamine-formaldehyde (MF) resins. Amino resins are considered to supplement and complement phenolic resins. UF and MF resins are prepared by reacting formaldehyde with urea and melamine, respectively. The general reaction scheme for the synthesis of UF and MF resins is shown in Figure 4.5. Liquid amino resins are prepared by reacting the corresponding methylol intermediate in n-butanol or methanol.

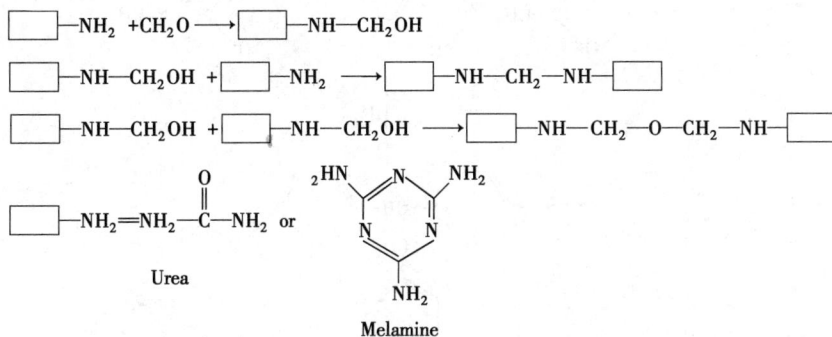

Figure 4.5   The general reaction schemes for synthesis of UF and MF resins

Amino resins are used as curing agents for the resins containing carboxyl, hydroxyl and amide groups. Amino resins can be used in all possible applications in which phenolic resins are used. Unlike phenolic resins, which are dark-brown in color, amino resins are light in color. Hence amino resins have replaced phenolic resins in applications where bright color and aesthetic appearances are needed. MF resin shows the best performance in terms of water resistance and toughness. However, use of MF resin is restricted due its high cost. The first commercial production of MF resin was started by American Cyanamid Corporation (USA) in 1939. The well-known applications of phenolic resins are in domestic plugs and switches. UF resins have largely replaced phenolic resins for such purposes due to their better anti-tracking properties and wider color range.

4.5   Furan resins

Furan resins are also a low-volume consumption resin like amino resins, and are used as supplements to phenolic resins[5]. They are prepared by the reaction between a phenol and furan compounds such as furfural, furfuryl alcohol, and furan. Furan compounds can be used in place of formaldehyde in the conventional production of phenolic resins. The most popular and viable furan resins are prepared from furfuryl alcohol (FFA). FFA undergoes homopolymerisation through an addition reaction in acid medium, leading to the formation of a furan resin (Figure 4.6). Like other

113

thermosetting resins, furan resin undergoes crosslinking in the presence of a strong acid and forms a 3D network. The crosslinking involves the reaction with a methylene bridge.

Like amino resins, furan resins can also be used in all possible applications where phenolic resins are used.

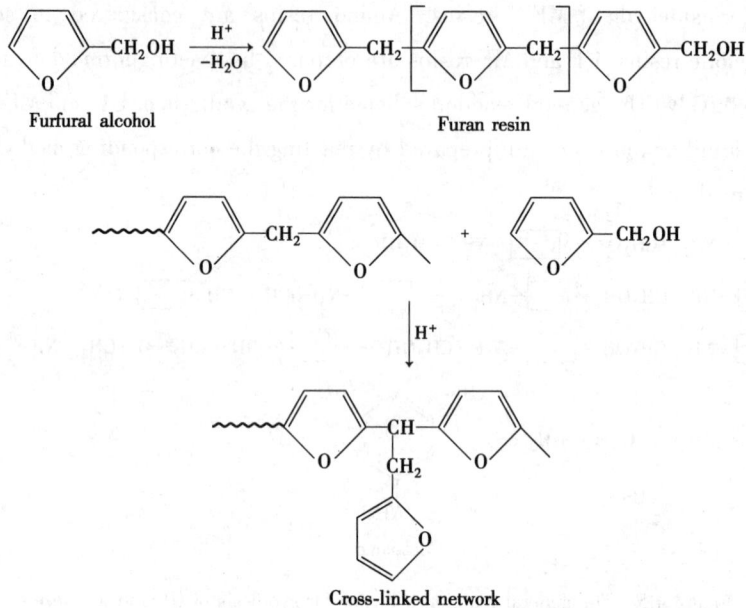

Figure 4.6　Reaction schemes for the synthesis and crosslinking of furan resin

4.6　Unsaturated polyester resins

The second synthetic thermoset resin discovered in early 1940 (after phenolic resin) was unsaturated polyester (UPE) resin. UPE consists of an unsaturated polyester, a monomer, and an inhibitor. UPE gained wide industrial applications due to their low viscosity, which offers easy processability, low cost and rapid cure schedules.

Polyesters are macromolecules made by reacting a diacid or dianhydride with a dihydroxy compound (diols). To make unsaturated polyesters, maleic anhydride or fumaric acid is used in addition to a saturated acid, which provides unsaturation in the structure. The most commonly used anhydrides are maleic anhydride (unsaturated) and phthalic anhydride (saturated). The commonest diols are ethylene glycol or propylene glycol. Use of an unsaturated anhydride is very critical to provide unsaturation in the structure, which is utilised to cure the resin by free-radical polymerisation. The chemical reaction for the synthesis of UPE is shown in Figure 4.7.

Fumarate double bonds (planar trans configuration) react faster with the reactive monomer compared with the double bonds of maleate ester because maleate esters are slightly distorted from

114

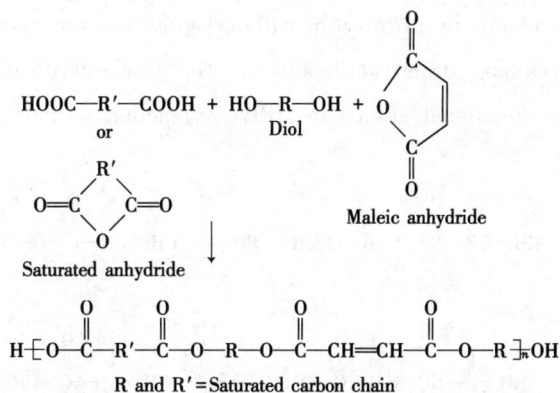

Figure 4.7   Reaction scheme for the synthesis of UPE resin

the planar configuration, which suppresses their ability to copolymerise with styrene. Hence fumaric acid is preferable to maleic anhydride as an unsaturated anhydride precursor for the formation of a uniform network with better properties. However, maleic anhydride is mostly used because of two reasons: ①maleic acid offers lower cost and easier handling, and ② most maleate double bonds isomerise to a more stable fumarate form during resin synthesis at high temperature[6]. It was reported that the extent of isomerisation depended on the structure of glycol. For instance, isomerisation is reported to be 95% with propylene glycol, 39% with 1,4-butylene glycol, and 35% with 1,6-hexamethylene glycol at 180 ℃. ^{13}C-NMR analysis is used to study the isomerisation. The nature of acids and catalyst also has an important role in isomerisation [7].

The rate of polyester synthesis by polycondensation depends on the chemical structure of the reactants (diacid and diol) and the stoichiometry of the reactants. In general, glycol is used in slight excess to compensate for the potential loss of glycol via evaporation. The reaction takes place in two stages: formation of monoester followed by polycondensation at higher temperature. The reaction is reversible and water is produced as a byproduct. Hence it is necessary to remove the water from the reaction mixture to push the reaction forward. Water is removed continuously from the reaction mixture by application of vacuum or by using a solvent such as xylene, which forms an azeotrope with water in the vapour state. A Dean and Stark apparatus is used for the synthesis. The azeotrope vapour is allowed to condense in a receiver tank, where they are separated from each other due to the difference in density. Xylene forms the upper layer (which is fed back continuously to the reactor) and water forms the bottom layer (which is removed through an opening at the bottom of the tank). The progress of the reaction is monitored through the amount of water produced and the acid value (mg of potassium hydroxide required to neutralise 100 g of resin) of the reaction mixture. The acid value is checked by withdrawing a small amount of sample from the reactor and it analysing using a standard titration method. The typical molecular weight of UPE is 3 000 ~ 5 000 g/mole with an acid value of the product of <20. Polymerisation is the first growth step, so the UPE of the polymer is highly sensitive to the purity of the reactants. At high temperature, alcoholysis or acidolysis of the polyester chains by hydroxyl or carboxyl groups of the monomers and/or oligomers occurs. This process is called "transesterification". The transesterification reaction allows the redistribution of UPE and functional groups in UPE. A reaction scheme for transesterification is shown in Figure 4.8.

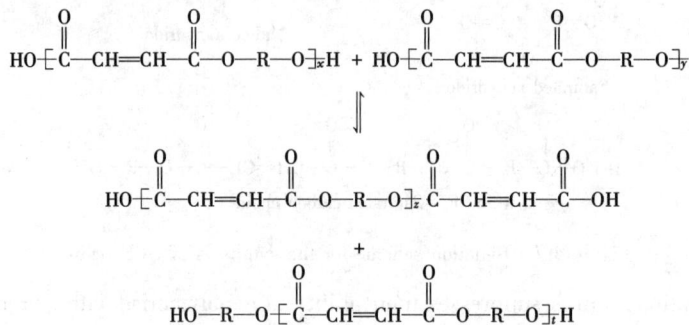

$$HO \left[\overset{O}{\overset{\|}{C}} - CH = CH - \overset{O}{\overset{\|}{C}} - O - R - O \right]_x H \; + \; HO \left[\overset{O}{\overset{\|}{C}} - CH = CH - \overset{O}{\overset{\|}{C}} - O - R - O \right]_y$$

$$\updownarrow$$

$$HO \left[\overset{O}{\overset{\|}{C}} - CH = CH - \overset{O}{\overset{\|}{C}} - O - R - O \right]_z \overset{O}{\overset{\|}{C}} - CH = CH - \overset{O}{\overset{\|}{C}} - OH$$

$$+$$

$$HO - R - O \left[\overset{O}{\overset{\|}{C}} - CH = CH - \overset{O}{\overset{\|}{C}} - O - R - O \right]_r H$$

Figure 4.8   Reaction scheme for a transesterification reaction

A linear structure of polyester is expected from the reaction of acid and alcohol (Figure 4.7). However, branching in a polyester structure takes place as a result of side reactions. Such side reactions were first investigated by Ordelt and co-workers. The electron-deficient double bond (due to the presence of electron-withdrawing carbonyl groups) of maleic anhydride can react with the hydroxyl groups of glycol or oligomers via Michael addition. The reaction of glycol with the double bonds produces short branches, whereas involvement of hydroxyl groups of oligomer or macromolecules leads to the formation of long branches (Figure 4.9).

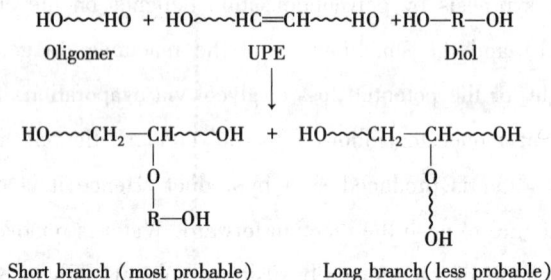

$$HO \sim\sim HO \; + \; HO \sim\sim HC = CH \sim\sim HO \; + HO - R - OH$$

Oligomer   UPE   Diol

$$\downarrow$$

$$HO \sim\sim CH_2 - \underset{\underset{R-OH}{\overset{|}{O}}}{CH} \sim\sim OH \; + \; HO \sim\sim CH_2 - \underset{\underset{OH}{\overset{|}{O}}}{CH} \sim\sim OH$$

Short branch (most probable)   Long branch(less probable)

Figure 4.9   Mechanism for the formation of long and short branches in UPE resin

The most important application of unsaturated polyester is to make the fiber reinforced plastics (FRP). The matrix is a continuous phase and the reinforcement is a discontinuous one. The duty of reinforcements is attaining strength of the composite and the matrix has the responsibility of bonding of the reinforcements. There are recognizable interface between the materials of matrix and reinforcements. The composite materials, however, generally possess combination of properties such as stiffness, strength, weight, high temperature performance, corrosion resistance, hardness and conductivity which are not possible with the individual components. Indeed, composites are produced when two or more materials or phases are used together to give a combination of properties that cannot be achieved otherwise. Composite materials especially the fiber reinforced polyester kind highlight how different materials can work in synergy. Analysis of these properties shows that they depend on ① the properties of the individual components; ② the relative amount of different phases; ③ the orientation of various components; the degree of bonding between the matrix and the reinforcements and ④ the size, shape and distribution of the discontinuous phase. The material involves can be organics, metals or ceramics. Therefore, a wide range of freedom exists, and composite materials can often be designed to meet a desired set of engineering properties and characteristics[8].

There are many types of composite materials and several methods of classifying them. One method is based on the matrix materials which include polymers, metals and ceramics. The other method is based on the reinforcement phase which has the shape of fiber, particulate and whisker. Whiskers are like fibers but their length is shorter. The bonding between the particles, fibers or whiskers and the matrix is also very important. In structural composites, polymeric molecules known as coupling agent are used. These molecules form bonds with the dispersed phase and become integrated into the continuous matrix phase as well. The most popular type of composite material is the fiber-reinforced polyester composites, in which continuous thin fibers of one material such as glass, carbon or natural fibers are embedded in a polyester matrix. They are also called glass fiber reinforced polyester (GFRP), carbon fiber reinforced polyester (CFRP) and natural fiber reinforced polyester (NFRP). The objective is usually to enhance strength, stiffness, fatigue, resistance, or strength to weight ratio by incorporating strong and stiff fibers in a softer, more ductile matrix. The microstructure of a selected GFRP composite is shown in Figure 4.10.

Figure 4.10　Microstructure of fiberglass reinforced polyester composite

Mixtures of chopped fiberglass and polyester prepolymers and fiberglass mat impregnated with polyester prepolymer are called bulk molding compounds (BMC) and sheet molding compounds (SMC), respectively.

117

About 80% of unsaturated polyesters (excluding alkyd resins) are used to produce reinforced products, including electrical, marine, and transportation applications. Speedboat and motorboat hulls are generally produced by the SMC process, in which a mixture of unsaturated polyester resin, fibers (often fiberglass), and fillers is held between sheets of polyethylene film until it thickens to a leathery sheet. These sheets are molded under pressure to give fiber-reinforced plastic hulls. Shower stalls and industrial tubs are also made by the SMC process.

4.7 Epoxy resins

Epoxy resins are a class of thermosetting resin materials characterised by two or more oxirane rings or epoxy groups within their molecular structure[9]. The commonest epoxy resin is the diglycidyl ether of bisphenol A (DGEBA), which is prepared by the reaction of epichlorohydrin (ECD) and bisphenol A (BPA) (Figure 4.11).

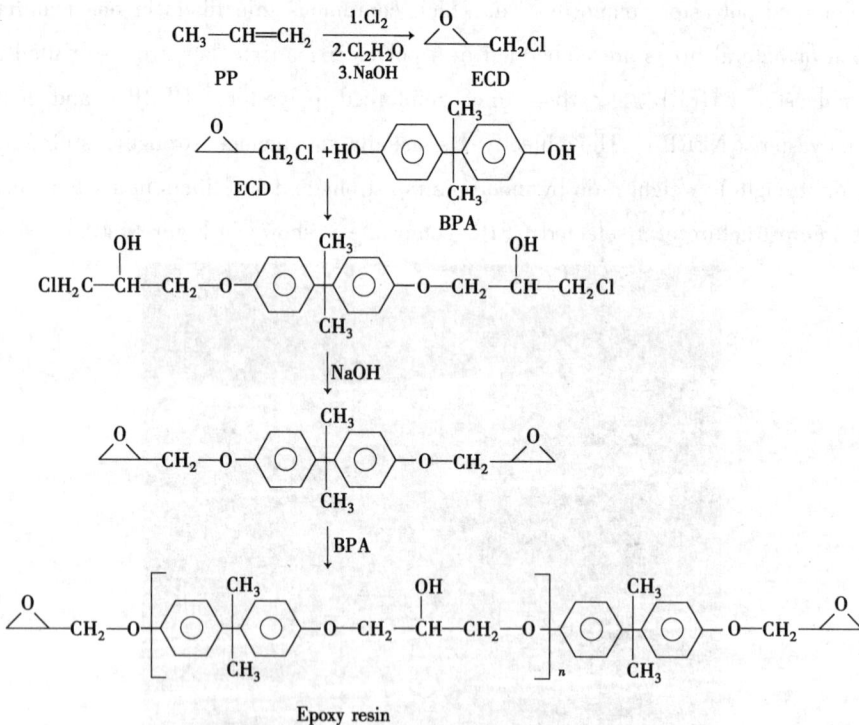

Figure 4.11　Reaction schemes for the synthesis of DGEBA-type epoxy resin

ECD is prepared from polypropylene (PP) by reacting chlorine with sodium hydroxide. ECD is allowed to react with BPA in the presence of sodium hydroxide. The first step is cleavage of the oxirane ring of ECD by the hydroxyl group of BPA. The second step is cyclisation in base medium, leading to the formation of an epoxy-ended intermediate. The intermediate then undergoes chain extension with BPA to produce an epoxy resin. A wide variety of resins can be produced by adjusting the concentration of the reactants. A liquid resin can be further chain-extended with BPA to make a solid resin of higher molecular weight. Thus epoxy resins are available in various consistencies from low viscous liquid to a tack-free solid.

There are undoubtedly more publications and reports based on the basic and applied research on epoxy resins than for any other commercially available thermosetting resin. The broad interest in epoxy resins originates from the versatility of epoxy group towards a wide variety of chemical reactions and the useful properties of the network polymers such as high strength, very low creep, excellent corrosion-and weather-resistance, elevated temperature service capability, and adequate electrical properties.

4.8 Polyurethanes[10]

PU is polymer containing urethane linkages and is available in thermoplastic and thermosetting forms. In 1935, Otto Bayer and coworkers invented PU as a product of the polyaddition reaction between a macroglycol and a diisocyanate. PU are made as castable PU elastomers, PU thermoplastic elastomers and PU engineering thermoplastics. The precursors of standard PU formulations are polyol, polyisocyanate, extender and modifier (which are optionally used).

4.8.1 Polyols

Polyols are hydroxyl-functionalized oligomers having a UPE in the range 300 ~ 9 000 g/mole and functionality of 1 ~ 6 equivalent per mole. Linear and low functionality ($f = 2 \sim 3$ eq/mole) generates flexible (low modulus) PU, whereas branched and high functionality ($f = 3 \sim 6$ eq/mole) polyols lead to hard PU systems (high modulus). Depending on the backbone structure, polyols are classified into two groups: polyether polyol and polyester polyol. Polyether polyol comprises about >80% of global PU. Polyether polyol is prepared by the addition reaction of epoxide with a molecule with active hydrogen or by ionic polymerisation of alkylene oxide.

Polyester polyols are produced by an esterification reaction of a carboxylic acid and a glycol. Unlike in chain-growth polymerisation, the UPE of the product of a step growth polymerisation is highly sensitive to the stoichiometry of the reactants. Thus high molecular weight linear polyester polyols are prepared using high-purity acids and are used to produce PU with improved properties, which are not achievable by using polyether polyol. Low molecular weight branched polyester polyols are prepared by glycolysis (trans esterification) of recycled byproducts with glycols. The various polyols used for PU synthesis are listed in Table 4.2. The chemical structures of some commonly used polyols are shown in Figure 4.12.

$$HO-(CH_2CH_2CH_2CH_2O)_{\overline{n}}H \qquad HO-(CH_2CH_2O)_{\overline{n}}H$$

$$(a)\,PTMO \qquad\qquad (b)\,PEO$$

$$HO-(CH_2CHO)_{\overline{n}}H \qquad HO-(CH_2)_5\overset{\displaystyle O}{\overset{\|}{C}}-O-(CH_2)_2OH$$
$$\underset{CH_3}{|}$$

$$(c)\,PPO \qquad\qquad (d)\,PCL\ diol$$

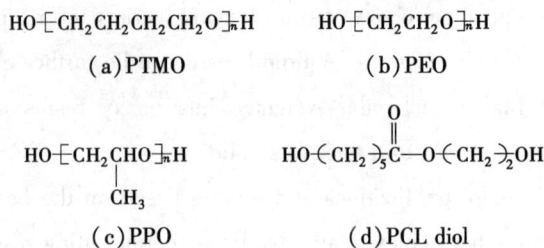

Figure 4.12 Chemical structures of polyols used for PU synthesis

4.8.2 Isocyanates

The second component for PU is a monomer containing two or more isocyanate functional groups. The isocyanate compounds commonly used for the synthesis of PU are listed in Table 4.2. In general, aromatic isocynates, especially 2,4-and 2,6-toluene diiso cyanate (TDI), 4,4'-diphenylmethane diisocyanate (MDI) and 1,6-hexamethylene diisocyanate (HDI) (Figure 4.13) are used for the synthesis of thermosetting PU. The high reactivity of isocyanates does not allow use as a one-component system. Isocyanates are highly toxic, which is a concern for storage of these materials. To circumvent this, they are reacted with phenols, oximes, alcohols, or dibutyl malonate; these substances are called "blocking agents". A blocking agent must be selected in such a way that the blocked isocyanate undergoes deblocking in the reaction condition and generates isocyanate, which reacts with polyol; or the blocking agent should be eliminated by the polyol during the reaction. The reaction between blocked isocyanate and the polyol is shown in Figure 4.14. The rate, extent and mechanism of reaction depend on the chemical nature of the blocking agent and polyol, catalyst, and polarity of the solvent.

Table 4.2 Polyol and isocyanate compounds commonly used for the synthesis of PU

Isocyanate	Polyol
4,4'-Diphenylmethane diisocyanate (MDI)	Poly(ethylene oxide) (PEO)
2,4-and 2,6-Toluene diisocyanate (TDI)	Poly (propylene oxide) (PPO)
1,6-Hexamethylene diisocyanate (HDI)	Poly (tetramethylene oxide) (PTMO)
1,5-Napthalene diisocyanate	Poly caprolactone (PCL) diol
4,4'-Dicyclohexyl methane diisocyanate	1,4 polybutadiene diol
3-Isocyanatomethyl-3,5,5-	Poly (ethylene adipate)
trimethylcyclohexyl isocyanate (isophorone diisocyanate)	Poly (dimethyl siloxane)
para-phenylene diisocyanate	Polyisobutylene diol
2,2,4-Trimethyl-1,6-hexamethylene	

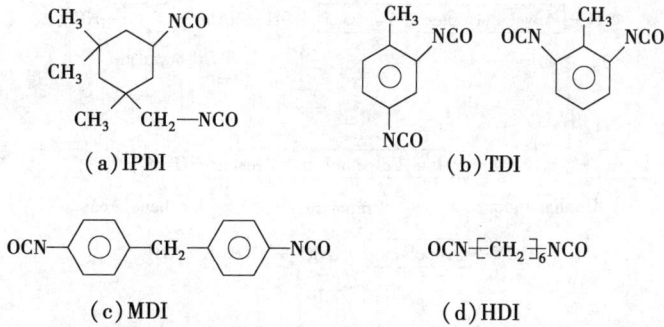

Figure 4.13   Chemical structures of common isocyanates used for PU synthesis

Blocking

$$RN=C=O + BA-H \longrightarrow R-NH-\overset{\overset{\displaystyle O}{\|}}{C}-BA$$

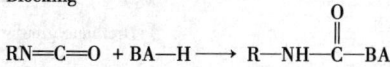

Deblocking and urethane formation

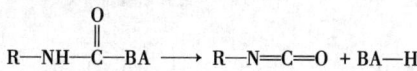

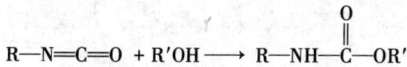

$$R-NH-\overset{\overset{\displaystyle O}{\|}}{C}-BA \longrightarrow R-N=C=O + BA-H$$

$$R-N=C=O + R'OH \longrightarrow R-NH-\overset{\overset{\displaystyle O}{\|}}{C}-OR'$$

Urethane formation by elimination-addition reaction

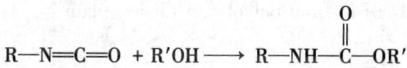

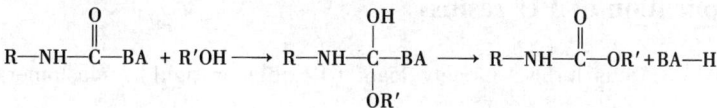

$$R-N=C=O + R'OH \longrightarrow R-NH-\overset{\overset{\displaystyle O}{\|}}{C}-OR'$$

$$R-NH-\overset{\overset{\displaystyle O}{\|}}{C}-BA + R'OH \longrightarrow R-NH-\overset{\overset{\displaystyle OH}{|}}{\underset{\underset{\displaystyle OR'}{|}}{C}}-BA \longrightarrow R-NH-\overset{\overset{\displaystyle O}{\|}}{C}-OR'+BA-H$$

BA = Blocking agent

Figure 4.14   The chemistry of blocking of isocyanate and reaction
of blocked isocyanate with the polyol

4.8.3   PrePolymers

The synthesis of PU is shown in Figure 4.15. The polyol is first reacted with excess di-or polyisocyanate to get an isocyanate-terminated intermediate, known as a "prepolymer". If all the hydroxyl groups are capped with isocynates and no free isocyanate remains in the mixture, then the intermediate is called a "full" prepolymer. Such prepolymers are formed if the isocyanate groups on the polyisocyanate have different reactivity, as in the case of 2,4-TDI and the ratio of equivalent of isocyanate to hydroxyl (f_{NCO}/f_{OH}) is close to 2. If the isocyanate groups are similar in reactivity or the functionality ratio (f_{NCO}/f_{OH}) is >2, then isocyanate groups will not be consumed fully and some amount of isocyanate will remain free. Prepolymers are analysed for isocyanate content using standard methods.

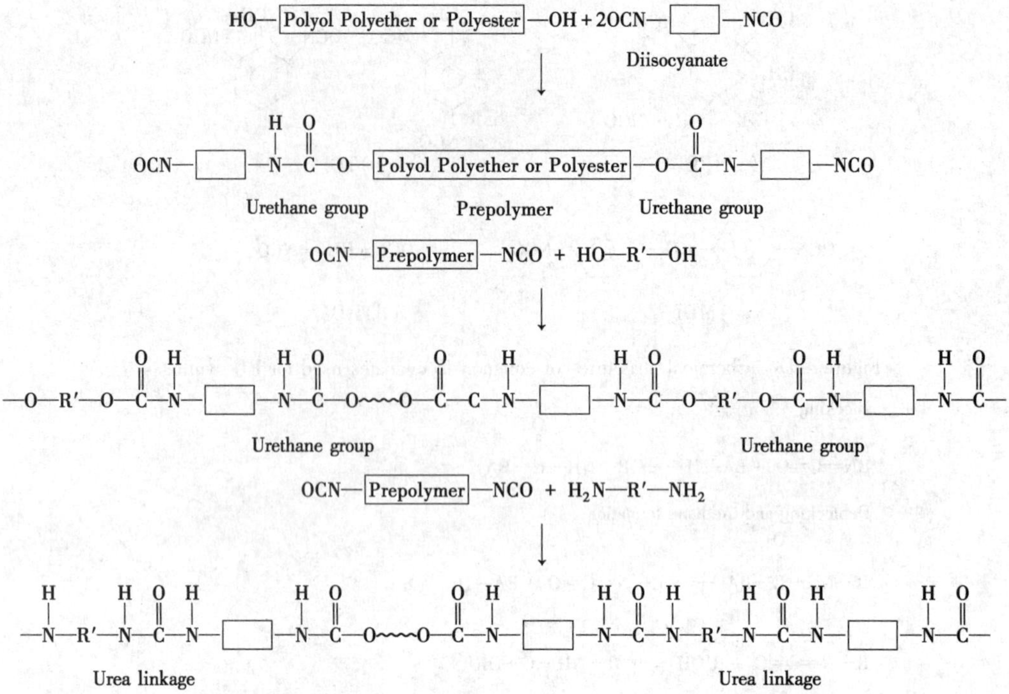

Figure 4.15   Synthesis of PU resin from polyol and diisocyanate

4.8.4 **Application of PU resins**

PU are used in various forms, namely foam (flexible or rigid), elastomer sheet, coating, adhesives and sealants.

The major use of rigid foams is for refrigeration insulation such as in domestic refrigerators and cold storage rooms. They are also used in building and construction industries. The examples are roof or wall insulation for domestic and industrial buildings. Rigid foams are also used for insulated trailers, trucks and railway cars. Flexible foams are used in furniture, cushioning for transportation and sitting and bedding applications because of their lightweight and excellent cushioning properties. Bedding applications include mattresses, topper pads, convertible sofas and mattresses of variable size and densities.

PU coatings are known for their excellent abrasion resistance and weather resistance. They are used for coating of equipment, textiles and leather. Because of their better water resistance compared with epoxy, they replace epoxy for surface coating and paint in marine industries. Automotive coating includes clear top coating, plastic parts coating and body primers. PU coatings are also used for coating military and civilian aircraft.

PU elastomers processed by RIM are widely used in automotive industries. Examples are bumper covers, external body panels, modular windows and exterior and interior trims. Such elastomers are also used in equipment housing, sports equipment and furniture. Castable PU elastomer sheets are used as vibration damping materials and acoustic window materials for various naval and civil applications. Because of higher water resistance, PU elastomers are preferable as encapsulant materials for underwater electronics.

Cellular castable PU elastomers are used extensively in footwear industries. The introduction of multicolour, multi-density microcellular PU products has further broadened their applications. Non-cellular products are widely used as bushings, gaskets, hoses, belts, shock damping mounts and moulded parts for automobiles. PU sealants are used for household appliances, toys, and in ships and submarines.

The largest consumer of PU adhesive is the textile industry. Applications include textile lamination, integral carpet manufacture and rebonding of foam. Rebonded foam is made using scrap PU foam bonded with a urethane prepolymer and used primarily as carpet underlay. PU adhesives are used to bond film to film, film to foil and film to paper in various packaging constructions. Other uses of PU adhesives are for laminating composite panels in truck and car applications, polycarbonate headlamp assemblies, and door panels. PU adhesives have replaced neoprene-based adhesive for footwear applications.

PU is potential materials for shape memory applications which have drawn considerable attention in recent years. Suitably designed PU exhibit an excellent shape memory effect [11]. A material is said to show shape memory effect if it can be deformed and fixed into a temporary shape and recover its original permanent shape only on exposure of external stimuli such as heat or light. Thermally induced shape memory effect is more common if the recovery takes place with respect to a certain critical temperature. The most widely used shape memory material is Ni-Ti alloy (Nitinol). Shape memory alloys (SMA) exhibit outstanding properties such as small size and high strength, and have found wide technical applications. However, they have obvious disadvantages such as high manufacturing cost, limited recoverable deformation and appreciable toxicity. PU offers deformation to a much higher degree and a wider scope of varying mechanical properties compared with SMA or ceramics, in addition to its inherent advantages of being cheap, lightweight and easy to process. In the case of SMA, the maximum recoverable strain is 8%, whereas in PU it can be 800%. PU are also biocompatible, non-toxic and can be made biodegradable.

Shape memory PU and polymers in general have tremendous applications in biology and medicine especially for biomedical devices which may permit new medical procedures. Because of the ability to memorise a permanent shape that can be substantially different from an initial temporary phase, a bulky device could be introduced into the body in a temporary shape (e.g., string) that could go through a small laparoscopic hole and then be expanded on demand into a permanent shape at body temperature.

Shape memory PU has been proposed as a candidate for aneurysm coils. An intracranial aneurysm can go undetected until the aneurysm ruptures, causing hemorrhaging within the subarachnoid space surrounding the brain. The typical treatment for large aneurysms is remobilisation using platinum coils. However, in about 15% of the cases treated by platinum coils, the aneurysm eventually re-opens as a result of the bio-inertness of platinum. One solution is to develop suitable materials with increased bio-activity (e.g., SMP) to use as coil implants.

Another example of a biomedical application is a microactuator made from thermosetting PU, which has been used to remove blood clots. A microactuator with a permanent shape of a cone-shaped coil can be elongated to a straight wire and fixed before surgery and delivered to an occlusion through a catheter. On triggering the shape recovery using an optical heating method, the original coil shape is recovered and blood flow restored.

Recently, the concept of cold hibernated elastic memory utilising SMP in open cellular structures was proposed for space-bound structural applications. The concept of cold-hibernated elastic memory can be extended to various new applications such as microfoldable vehicles, shape determination and microtags. Recent studies on shape memory PU-based conductive composites using conducting polymers and carbon nanotubes show considerable promise for application as electroactive and remote sensing actuators.

Collection of Exercises

1. List the main applications of epoxy plastics.
2. What are the main processing methods of thermosetting plastics?
3. Compared with thermoplastics, what are the advantage and disadvantage of thermosetting plastics?
4. What are meant by A-stage and B-stage phenolics?
5. What are the characteristics and application categories of amino plastics
6. List the main applications of unsaturated polyester.
7. List the properties of glass fiber (GF) reinforced unsaturated polyester and GF reinforced epoxy plastics.
8. Which is the faster reaction: step reaction or chain reaction polymerization?
9. Write down the formulations of epoxy resin.
10. Which of the following are thermoset plastics: Melamine dishware, Bakelite, hard rubber?

REFERENCES

[1] F. Reed Estabrook, Alan Low, Tool and Manufacturing Engineers Handbook (Chapter 13

Compression and Transfer molding), Society of Manufacturing Engineers, 1998.

［2］http://www.plenco.com/plenco_processing_guide/, Plenco 公司技术资料.

［3］Peter W. Kopf, Encyclopedia of Polymer Science and Technology (Phenolic Resins), Wiley, 2002.

［4］Laurence L. Williams, Encyclopedia of Polymer Science and Technology (Amino Resins), Wiley, 2002.

［5］Debdatta Ratna, Handbook of Thermoset Resins, Smithers Rapra, Shawbury, Shrewsbury, Shropshire, United Kingdom, 2009.

［6］L.V. Cristobal and G.A.P Mendoza, Unsaturated polyesters, Polymer Bulletin, 1989, 22(5-6), 513-519.

［7］M. Malik, V. Choudhary, I. K. Varma, Current Status of Unsaturated Polyester Resins, Journal of Macromolecular Science, Part C: Polymer Reviews, 2000, 40(2-3): 139-165.

［8］http://www.intechopen.com/books/polyester, Salar Bagherpour, Fibre Reinforced Polyester Composites (chapter 6), InTech co., 2012.

［9］Eds., H.F. Mark and J.I. Kroscwitz, Encyclopedia of Polymer Science and Engineering, Volume 6, 2nd Edition, Wiley, New York, NY, USA, 1991, p.322.

［10］Ed., S.H. Goodman, Handbook of Thermoset Plastics, 2nd Edition, Noyes Publications, Park Ridge, New Jersey, USA, 1998.

［11］Linda Domeier, April Nissen, Steven Goods, et al. Thermomechanical Characterization of Thermoset Urethane Shape-Memory Polymer Foams, Journal of Applied Polymer Science, 115: 3217-3229, 2010.

CHAPTER 5

FIBERS

5.1 Introduction

Textile fiber can be divided into two categories: natural fiber and synthetic fiber[1], which is fiber chemical processed from natural or synthetic polymer compounds.

①Natural fiber: Natural fiber is that which is produced naturally. The source of origin could be vegetable, animal and mineral origin. The main representative products are: bast fibers, leaf fibers, seed and fruit fibers (such as cotton, coir), wool, silk, asbestos and so on.

②Synthetic fiber: Fiber which takes oil, natural gas, coal and agricultural and sideline products as material and then become synthetic macromolecular compound by a series of chemical reaction and finally made by machining. The main representative products are: polyester, polyamide (nylon), polyolefin fiber (polypropylene and polythene, etc.), polyacrylonitrile fiber (acrylic), poly vinyl alcohol fiber (whalen), polyvinyl chloride fiber (chloro fiber) and so on.

Other than the six kinds of synthetic fibers above, there are else fibers that own good properties, such as: aromatic polyamide (aramid fiber), aromatic polyester, etc.

The classification of textile fiber can be summarized in the Figure 5.1.

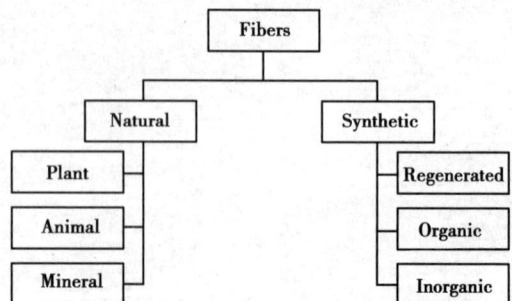

Figure 5.1   Classification of fibers

5.2　Natural fibers

5.2.1　Cotton [2]

Cotton is a natural fiber and makes up just under half of all the fiber sold in the world. Cotton grows on a plant that is a member of the Hibiscus family and is botanically known as *Gossypium hirsutum* or *barbadense*. By nature it is a perennial shrub which reaches a height of 3.5 m. Commercially it is grown as an annual and only reaches a height of 1.2 m (Figure 5.2).

Figure 5.2　The cotton plant

The most common type of cotton grown in Australia is *Gossypium hirsutum*, more commonly known as American Upland. It is a leafy, green shrub that briefly has cream and pink flowers that become the "fruit" or cotton bolls.

The cotton plant has a deep taproot, which can go as deep 1.5 m. It is fairly drought-tolerant but requires a regular and adequate moisture supply to produce profitable yields.

Cotton today is the most used textile fiber in the world. Its current market share is 56% for all fibers used for apparel and home furnishings and sold in the U.S.[3]. Another contribution is attributed to nonwoven textiles and personal care items. It is generally recognized that most consumers prefer cotton personal care items to those containing synthetic fibers. World textile fiber consumption in 1998 was approximately 45 million tons. Of this total, cotton represented approximately 20 million tons. The earliest evidence of using cotton is from India and the date assigned to this fabric is 3000 B.C. There were also excavations of cotton fabrics of comparable age in Southern America. Cotton cultivation first spread from India to Egypt, China and the South Pacific. Even though cotton fiber had been known already in Southern America, the large-scale cotton cultivation in Northern America began in the 16th century with the arrival of colonists to southern parts of today's United States. The largest rise in cotton production is connected with the invention of the saw-tooth cotton gin by Eli Whitney in 1793. With this new technology, it was possible to produce more cotton fiber, which resulted in big changes in the spinning and weaving industry.

Cotton is grown so the fiber can be made into products we use each day, including jeans, T-shirts, sheets and towels. Fiber from the cotton plant is made into yarn and fabric, the seeds are crushed for oil and animal feed, and the leaves are turned into mulch.

In its unprocessed form, the fiber is called lint. The lint grows inside the fruit of the cotton plant (the boll, Figure 5.3). Inside each boll are about 30 cotton seeds with many lint fibers attached to each seed. The lint is protected in the boll until it ripens and splits open.

Cotton, as a natural cellulosic fiber, has a lot of characteristics, such as: comfortable soft hand, good absorbency, color retention, prints well, machine-washable, dry-cleanable, good strength, drapes well, easy to handle and sew. Cotton is the shortest commercial textile fiber (Figure 5.4). Australian cottons are typically up to three centimetres long if irrigated, but shorter from dryland crops. The fiber is made of cellulose, which has a thin coating of wax. Cotton is an unusual fibre as it is a thin, hollow tube like a straw. In contrast, wool is a solid fibre covered in scales.

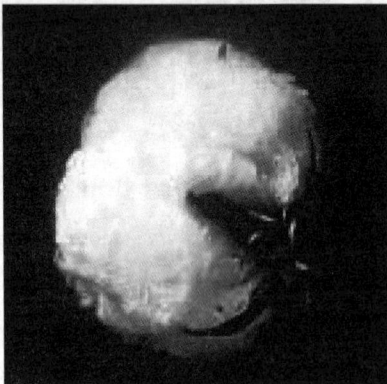

Figure 5.3 A ripe cotton boll

Figure 5.4 Rolls of cotton fabric

When cotton ripens the fibre collapses into a thin, twisted ribbon. The natural twist in the fibre allows cotton to be spun into yarn. Wool's spinability is due to the scales on the wool attaching to each other.

Natural fibres can be produced by plants or animals. Other fibre-producing plants include flax, hemp, sisal and jute. Natural animal fibres include wool from sheep, cashmere from goats and silk from silkworms.

Another group of fibres is the manufactured fibres like nylon and polyester. Natural and manufactured fibres are often blended together to make more versatile fabrics.

Almost every type of fabric available can be made with cotton fibers. The challenge is selecting the right fabric for the project. Lightweight cottons are best for shirts and dresses; medium-weight fabrics are suitable for pants, skirts, shirts, dresses, curtains, sheets and children's clothes; heavier fabrics are used for pants, outerwear, window treatments and work clothes.

Cotton fibers don't shrink, but cotton fabric does, so preshrink the yardage. To preshrink, wash the fabric the same way you intend to launder the finished garment.

Make sure the fabric is on-grain; that is, that the crosswise and lengthwise threads are truly perpendicular to each other. If the cotton has a permanent finish, it's not possible to straighten the grain. If the fabric has a print and the grain is off, the print may be skewed once you straighten the fabric. Avoid print fabrics unless the threads are truly on-grain.

There are no hard-and-fast rules for sewing with cotton because there so many fabric types. If the fabric does require special sewing techniques, it's because of the fabric type, not the fiber. Refer to fabric characteristics, such as ribbed or napped construction, decorative surfaces, loose weaves and fabric weight for sewing suggestions.

Most cotton fabrics can be laundered in the washing machine. They should be washed frequently, since they tend to absorb moisture and pick up dirt. Wash white items in hot water, medium colors in warm water and dark colors in cold water. Cotton will shrink more in hot water than cold, and fabric that's loosely woven shrinks more than tighter weaves.

Items with embossed designs and inner construction, such as a lining and shoulder pads, should be dry-cleaned. Loose knits, lingerie and fabrics with special finishes might also benefit from dry-cleaning.

5.2.2　Wool [4]

Warm and wonderful wool is nice to wear and easy to cut and sew. It's also one of the oldest known fibers, dating back to 4000 B.C. in ancient Mesopotamia and Babylonia. Today, wool is widely available and considered one of the most versatile fabrics. It's available in a wide range of weaves, blends, weights, textures and grades.

(1) What is wool?

Wool is derived from the fleece of many sheep breeds, each offering unique characteristics. It's a protein fiber like human hair, and has three layers responsible for wool's appealing attributes. The outer layer, called the epidermis, is made of overlapping scales with a coating that creates the ability to repel liquid, but absorb and evaporate moisture. The center (cortex) is comprised of long flat cells responsible for the fiber's natural crimp and elasticity, resulting in its inherent ability to resist wrinkles; higher quality fibers have more crimp than those of lesser quality. The innermost layer (medulla) varies in size and determines the fiber diameter; if it's fine, the fiber will be easy to spin and dye. A thick medulla results in a coarse fiber that is stiffer.

After sheep are sheared, the fleece is graded and sorted by length, fineness and color. Longer fibers are used for worsted yarns and shorter fibers are for woolen yarns. After sorting, the fibers are cleaned and may be dyed, but they can also be dyed after spinning or weaving. The wool is then carded—a process that separates the fibers into a smooth, fine web twisted to make roving. For smoother worsted fabrics, the roving is combed to remove shorter fibers. Woolen and worsted roving are both then spun into strands that are twisted into yarn (Figure 5.5).

(a) Fiber crimp makes for fuzzy wool fabrics   (b) Microscopic view of wool fibers

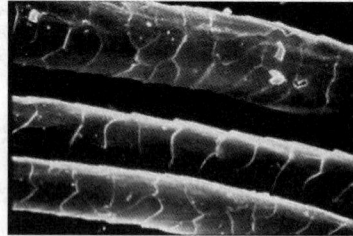

Figure 5.5   Photos courtesy of American Sheep Industry

(2) Woolen vs. worsted

Made from shorter fibers, yarns used for woolens are fuzzier, thicker and weaker than worsted yarns and are spun with a low to medium twist. They are used to make heavier fabrics with a slightly fuzzy surface, such as coatings, tweeds and flannels and are ideal for jackets, coats, skirts, blankets and rugs. Woolen fabrics are usually less expensive, more durable and felt more easily than worsted wool fabrics. They also hide stains better and are ideal for beginning sewers as stitching irregularities are more easily hidden in the fabric thickness.

(3) Wonderful wool characteristics

①Wicks moisture away from the body and evaporates it into the air.

②Water-resistant and repels light rain or snow.

③Insulates and keeps body heat in.

④Durable, resilient and resists abrasion and tearing, making it wear well and retain its shape. Adding a lining also helps retain shape.

⑤Resists wrinkling. Wrinkles more and is weaker when wet. Wrinkles are easily removed with steam.

⑥High-quality wools are less likely to feel scratchy when worn.

⑦Dyes easily and resists fading, but may fade in direct sunlight.

⑧Resists stains and cleans well. Because it shrinks and felts when exposed to heat and agitation, dry-cleaning is best unless the fabric is labeled "washable wool".

⑨Shapes easily with steam, making it easy to ease seams without puckering and shape curved areas.

⑩Can be damaged by hot iron, alkali-based stains and moths.

(4) Selecting fabric

Consider the garment style and the differences between worsted and woolen fabrics when selecting wools. For a dressy or tailored garment, or style with sharp creases, pleats or fine details, worsted wools are best. For casual wear with a heavier, textured look and less tailoring or soft gathers, woolens are best. Medium-weight wools are often easier to work with than very heavy or lightweight fabrics.

Examine the fabric to determine the quality. Closely-woven fabrics won't ravel easily when cut, resist snags and are more durable than those with a loose weave. Rub the fabric and squeeze it tightly in your hand, then let go. A higher quality fabric won't pill when rubbed and the wrinkles will fall out.

(5) Preparing fabric

Unless the fabric is labeled "pre-shrunk" or "ready-for-the-needle", test it first to determine if it's going to shrink. To test, cover a corner of the fabric where the selvedge and cut edge meet with a press cloth. Set a steam iron on the wool setting and press the iron in place for 10 seconds. Lift the iron; if the fabric has shrunken around the imprint of the iron, it should be pre-shrunk before cutting. Ask the dry cleaner to steam press it, or do it yourself.

Use a "with nap" layout when laying out pattern pieces; the scales of the wool fibers will sometimes reflect the appearance of the shading. For heavy or bulky wools, lay the pattern pieces out and cut one layer at a time to prevent distorting the shape of the pieces. For lightweight fabrics, pin closely to prevent the fabric from shifting as you cut. Use tailor's chalk or basting to mark darts and other pattern details. If the fabric appears the same on both sides, determine which side to use for the right side and mark the wrong sides of each piece after cutting.

Any interfacing can be used with wool fabric; use the fabric weight and drape as a guide when selecting interfacing. Always test fusible interfacing before using it to make sure it doesn't cause the surface of lightweight wools to pucker; if so, replace with sew-in interfacing.

(6) Sewing

To sew wool, use a universal needle size appropriate for the fabric weight: 60/8 for lightweight, 70/10 to 80/12 for medium weight or 90/14 for heavy weights. Use all-purpose thread, a standard presser foot and stitch length appropriate for the fabric weight.

Stay stitch bias and curved edges to prevent stretching. When sewing, press seams open to reduce bulk, grading them for heavy or bulky fabrics. To grade a seam allowance, trim each seam allowance a little narrower than the one above it, keeping the layer longest closest to the garment outside. Lining will help the garment keep its shape and eliminates the need for seam finishing. If the project is unlined, finish the seam allowances with serging, zigzagging or binding.

For best results when hemming your wool garment, hang it up for 24 hours before you mark the hemline.

(7) Take care

Always use a steam iron with a dry press cloth or a dry iron with a damp press cloth to press wool; dry heat will dry out the fibers. Let the fabric air-dry before folding or hanging to avoid wrinkles or stretching. To prevent a shine on dark or lightweight worsted wools, press from the wrong side; strips of brown paper under darts and seam allowances will prevent imprints. A tailor's ham is helpful for pressing curved areas, such as sleeve caps.

Dry-clean wool garments unless the fabric was labeled "washable".

5.3 Synthetic fibers

There are 2 types of synthetic fiber products, the semisynthetics, or cellulosics (viscose rayon and cellulose acetate), and the true synthetics, or noncellulosics (polyester, nylon, acrylic and modacrylic, and polyolefin). These 6 fiber types compose over 99 percent of the total production of manmade fibers in the U. S.

5.3.1 Manufacturing technologies for synthetic fibers [5]

Semisynthetics are formed from natural polymeric materials such as cellulose. True synthetics are products of the polymerization of smaller chemical units into long-chain molecular polymers. Fibers are formed by forcing a viscous fluid or solution of the polymer through the small orifices of a spinnerette (see Figure 5.6) and immediately solidifying or precipitating the resulting filaments. This prepared polymer may also be used in the manufacture of other nonfiber products such as the enormous number of extruded plastic and synthetic rubber products.

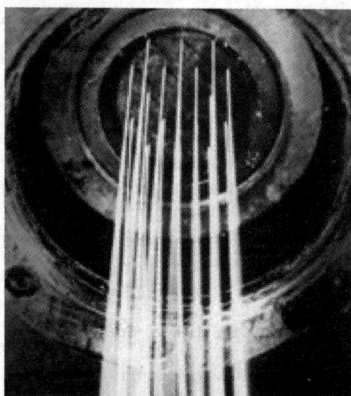

Figure 5.6　Spinnerette

Synthetic fibers (both semisynthetic and true synthetic) are produced typically by 2 easily distinguishable methods, melt spinning and solvent spinning. Melt spinning processes use heat to melt the fiber polymer to a viscosity suitable for extrusion through the spinnerette. Solvent spinning processes use large amounts of organic solvents, which usually are recovered for economic reasons, to dissolve the fiber polymer into a fluid polymer solution suitable for extrusion through a spinnerette. The major solvent spinning operations are dry spinning and wet spinning. A third method, reaction spinning, is also used, but to a much lesser extent. Reaction spinning processes involve the formation of filaments from prepolymers and monomers that are further polymerized and cross-linked after the filament is formed.

The spinning process used for a particular polymer is determined by the polymer's melting point, melt stability, and solubility in organic and/or inorganic (salt) solvents. (The polymerization of the fiber polymer is typically carried out at the same facility that produces the fiber.)

Table 5.1 lists the different types of spinning methods with the fiber types produced by each method. After the fiber is spun, it may undergo one or more different processing treatments to meet the required physical or handling properties. Such processing treatments include drawing, lubrication, crimping, heat setting, cutting, and twisting. The finished fiber product may be classified as tow, staple, or continuous filament yarn.

Table 5.1   Types of spinning methods and fiber types produced

Spinning Method	Fiber Type
Melt spinning	Polyester
	Nylon 6
	Nylon 66
	Polyolefin
Solvent spinning	Cellulose acetate
Dry solvent spinning	Cellulose triacetate
	Acrylic
	Modacrylic
	Vinyon
	Spandex
Wet solvent spinning	Acrylic
	Modacrylic
Reaction spinning	Spandex
	Rayon (viscose process)

(1) Melt spinning

Melt spinning uses heat to melt the polymer to a viscosity suitable for extrusion. This type of spinning is used for polymers that are not decomposed or degraded by the temperatures necessary for extrusion. Polymer chips may be melted by a number of methods. The trend is toward melting and immediate extrusion of the polymer chips in an electrically heated screw extruder. Alternatively, the molten polymer is processed in an inert gas atmosphere, usually nitrogen, and is metered through a precisely machined gear pump to a filter assembly consisting of a series of metal gauges interspersed in layers of graded sand. The molten polymer is extruded at high pressure and constant rate through a spinnerette into a relatively cooler air stream that solidifies the filaments(Figure 5.7). Lubricants and finishing oils are applied to the fibers in the spin cell. At the base of the spin cell, a thread guide converges the individual filaments to produce a continuous filament yarn, or a spun yarn, that typically is composed of between 15 and 100 filaments. Once formed, the filament yarn either is immediately wound onto bobbins or is further treated for certain desired characteristics or end use.

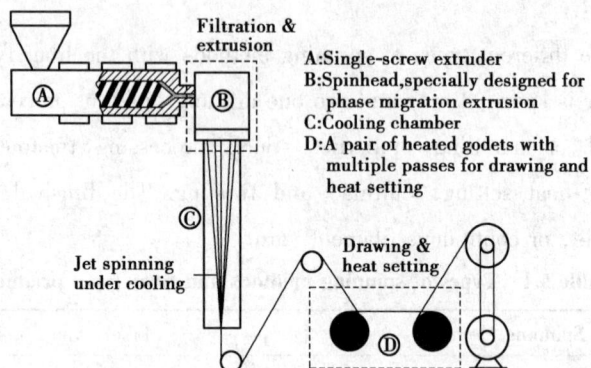

Figure 5.7　A typical melt spinning setup

Since melt spinning does not require the use of solvents, VOC emissions are significantly lower than those from dry and wet solvent spinning processes. Lubricants and oils are sometimes added during the spinning of the fibers to provide certain properties necessary for subsequent operations such as lubrication and static suppression. These lubricants and oils vaporize, condense, and then coalesce as aerosols primarily from the spinning operation, although certain post-spinning operations may also give rise to these aerosol emissions. Treatments include drawing, lubrication, crimping, heat setting, cutting, and twisting.

(2) Dry solvent spinning

The dry spinning process begins by dissolving the polymer in an organic solvent. This solution is blended with additives and is filtered to produce a viscous polymer solution, referred to as "dope", for spinning. The polymer solution is then extruded through a spinnerette as filaments into a zone of heated gas or vapor. The solvent evaporates into the gas stream and leaves solidified filaments, which are further treated using one or more of the processes described in the general process description section (Figure 5.8). This type of spinning is used for easily dissolved polymers such as cellulose acetate, acrylics, and modacrylics.

Dry spinning is the fiber formation process potentially emitting the largest amounts of VOCs per pound of fiber produced. Air pollutant emissions include volatilized residual monomer, organic solvents, additives, and other organic compounds used in fiber processing. Unrecovered solvent constitutes the major substance. The largest amounts of unrecovered solvent are emitted from the fiber spinning step and drying the fiber. Other emission sources include dope preparation (dissolving the polymer, blending the spinning solution, and filtering the dope), fiber processing (drawing, washing, and crimping), and solvent recovery.

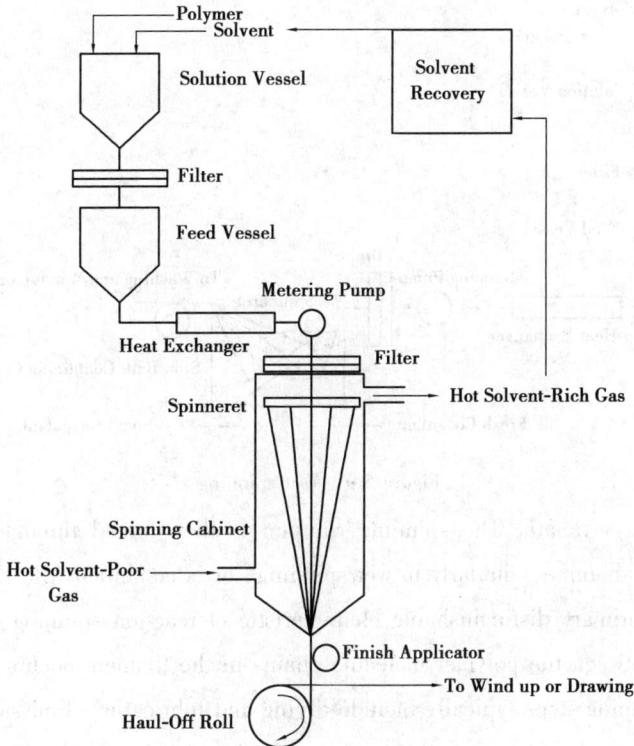

Figure 5.8   Dry spinning

(3) Wet solvent spinning

Wet spinning also uses solvent to dissolve the polymer to prepare the spinning dope. The process begins by dissolving polymer chips in a suitable organic solvent, such as dimethylformamide (DMF), dimethylacetamide (DMAc), or acetone, as in dry spinning; or in a weak inorganic acid, such as zinc chloride or aqueous sodium thiocyanate. In wet spinning, the spinning solution is extruded through spinnerettes into a precipitation bath that contains a coagulant (or precipitant) such as aqueous DMAc or water. Precipitation or coagulation occurs by diffusion of the solvent out of the thread and by diffusion of the coagulant into the thread. Wet spun filaments also undergo one or more of the additional treatment processes described earlier, as depicted in Figure 5.9.

Air pollution emission points in the wet spinning organic solvent process are similar to those of dry spinning. Wet spinning processes that use solutions of acids or salts to dissolve the polymer chips emit no solvent VOC, only unreacted monomer, and are, therefore, relatively clean from an air pollution standpoint. For those that require solvent, emissions occur as solvent evaporates from the spinning bath and from the fiber in post-spinning operations.

(4) Reaction spinning

As in the wet and dry spinning processes, the reaction spinning process begins with the preparation of a viscous spinning solution, which is prepared by dissolving a low molecular weight polymer, such as polyester for the production of spandex fibers, in a suitable solvent and a

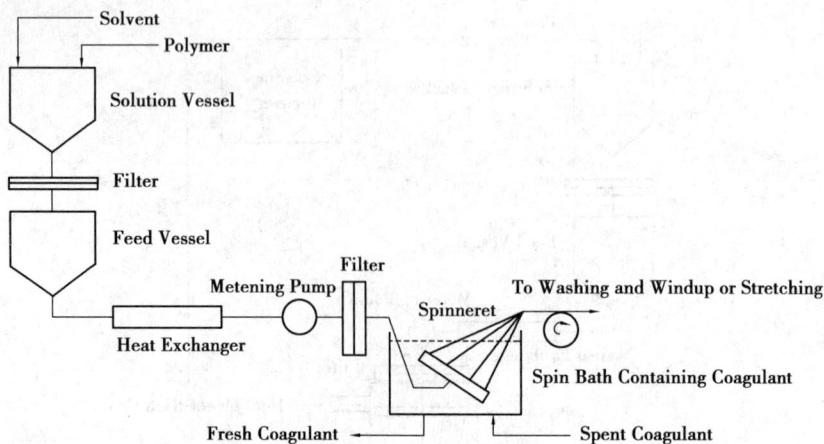

Figure 5.9   Wet spinning

reactant, such as diisocyanate. The spinning solution is then forced through spinnerettes into a solution containing a diamine, similarly to wet spinning, or is combined with the third reactant and then dry spun. The primary distinguishable characteristic of reaction spinning processes is that the final cross-linking between the polymer molecule chains in the filament occurs after the fibers have been spun. Post-spinning steps typically include drying and lubrication. Emissions from the wet and dry reaction spinning processes are similar to those of solvent wet and dry spinning, respectively.

(5) Emissions and controls

For each pound of fiber produced with the organic solvent spinning processes, a pound of polymer is dissolved in about 3 pounds of solvent. Because of the economic value of the large amounts of solvent used, capture and recovery of these solvents is an integral portion of the solvent spinning processes. At present, 94 to 98 percent of the solvents used in these fiber formation processes are recovered. In both dry and wet spinning processes, capture systems with subsequent solvent recovery are applied most frequently to the fiber spinning operation alone, because the emission stream from the spinning operation contains the highest concentration of solvent and, therefore, possesses the greatest potential for efficient and economic solvent recovery. Recovery systems used include gas adsorption, gas absorption, condensation, and distillation and are specific to a particular fiber type or spinning method.

The majority of VOC emissions from pre-spinning (dope preparation, for example) and post-spinning (washing, drawing, crimping, etc.) operations typically are not recovered for reuse. In many instances, emissions from these operations are captured by hoods or complete enclosures to prevent worker exposure to solvent vapors and unreacted monomer. Although already captured, the quantities of solvent released from these operations are typically much smaller than those released during the spinning operation. The relatively high air flow rates required in order to reduce solvent and monomer concentrations around the process line to acceptable health and safety limits make recovery economically unattractive. Solvent recovery, therefore, is usually not attempted.

The emission factors address emissions only from the spinning and post-spinning operations and the associated recovery or control systems. Emissions from the polymerization of the fiber polymer and from the preparation of the fiber polymer for spinning are not included in these emission factors.

Examination of VOC pollutant emissions from the synthetic fibers industry has recently concentrated on those fiber production processes that use an organic solvent to dissolve the polymer for extrusion or that use an organic solvent in some other way during the filament forming step. Such processes, while representing only about 20 percent of total industry production, do generate about 94 percent of total industry VOC emissions. Particulate emissions from fiber plants are relatively low, at least an order of magnitude lower than the solvent VOC emissions.

5.3.2　Rayon fiber [6]

Most rayon is made by the viscose process. The raw materials used in this process are cellulose wood pulp sheets or cotton linters, sodium hydroxide, carbon disulfide, and sulfuric acid. These are placed in a steeping press with contact in aqueous NaOH Solution for a period of $2 \sim 4$ hours at normal room temperature. A hydraulic ram presses out the excess alkali and the sheets are shredded to crumbs and aged for $2 \sim 3$ days. The aging process has its direct consequence on the viscosity of the solution.

Later CS_2 is added in a rotating drum mixer over a period of 3 hours. The orange cellulose xanthate which forms is transferred to a solubilizer, where in it is mixed into dilute caustic. The mixing of cellulose xanthate and dilute caustic yields a orange colored viscous solution which contains $7\% \sim 8\%$ cellulose and $6.5\% \sim 7\%$ NaOH. It is digested at room temperature for $4 \sim 5$ days. Thereafter, the solution is filtered and fed to spinning machines.

The spinning is carried out in extrusion spinnerettes, which are made up of platinum or gold alloys. The orifices of these spinnerets have a diameter of $0.1 \sim 0.2$ mm. for continuous filament yarns and of diameter $0.05 \sim 0.1$ mm for short fibre shapes. The solution extruded from the spinnerette is contacted with an acid bath which precipitates the filaments without causing them to break or stick together. The processing treatments of washing, desulphurizing, bleaching and conditioning takes place continuously and in order after the filaments are wound on a series of plastic rolls.

As shown in Figure 5.10, the series of chemical reactions in the viscose process used to make rayon consists of the following stages:

①Wood cellulose and a concentrated solution of sodium hydroxide react to form soda cellulose.

②The soda cellulose reacts with carbon disulfide to form sodium cellulose xanthate.

③The sodium cellulose xanthate is dissolved in a dilute solution of sodium hydroxide to give a viscose solution.

④The solution is ripened or aged to complete the reaction.

⑤The viscose solution is extruded through spinnerettes into dilute sulfuric acid, which regenerates the cellulose in the form of continuous filaments.

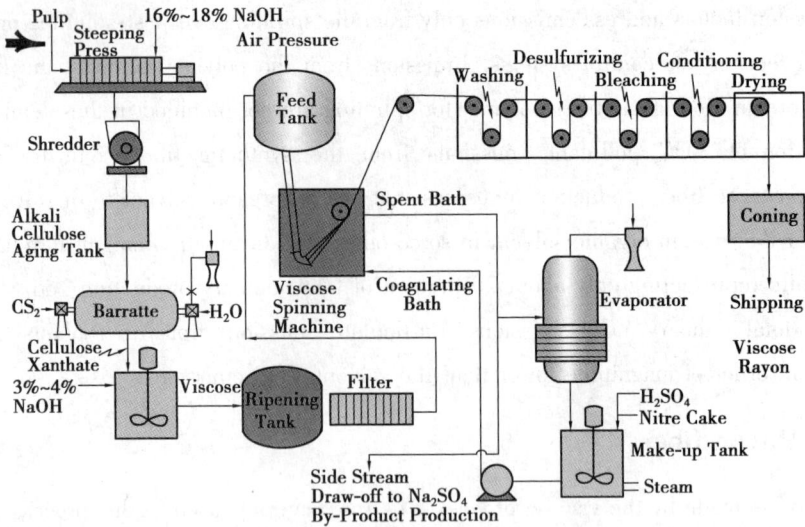

Figure 5.10 Rayon viscose process

Rayon fiber characteristics includes: highly absorbent; soft and comfortable; easy to dye; drapes well [7-8].

The drawing process applied in spinning may be adjusted to produce rayon fibers of extra strength and reduced elongation. Such fibers are designated as high tenacity rayon, which has about twice the strength and two-thirds of the stretch of regular rayon. An intermediate grade, known as medium tenacity rayon, is also made. Its strength and stretch characteristics fall midway between those of high tenacity and regular rayon.

Some major uses for rayon fiber are listed below:

①Apparel: Accessories, blouses, dresses, jackets, lingerie, linings, millinery, slacks, sport-shirts, sportswear, suits, ties, work clothes.

②Home Furnishings: Bedspreads, blankets, curtains, draperies, sheets, slipcovers, tablecloths, upholstery.

③Industrial Uses: Industrial products, medical surgical products, nonwoven products, tire cord.

④Other Uses: Feminine hygiene products.

5.3.3 Cellulose acetate and triacetate fiber[9]

Cellulose acetate fibers are consumed in two applications; the production of cigarette filter tow and textiles, mainly apparel. Cigarette filter tow accounted for approximately 92% of the world consumption of cellulose acetate fibers in 2011 and is expected to grow at an average annual rate of almost 2.5% during 2011—2016.

Textile applications accounted for nearly 8% of world consumption of cellulose acetate fibers in 2011; demand growth during 2011—2016 is forecast at an average annual rate of approximately

1.2%.

During 2008—2011, declines in demand for cigarette filter tow in North America, Western Europe and Japan were largely offset by consumption gains in Asia, mainly China.

The global economic slowdown, which began in 2008, reduced demand for cellulose acetate textile fiber as most consumption is in apparel and home furnishings. Since 2008, several producers have exited the market; this greatly increased global capacity utilization for cellulose acetate textile fibers, from 57% in 2008 to 89% in 2011. World demand is expected to grow moderately during 2012—2016, as most replacement by other textile fibers, especially polyester, has largely occurred and as demand recovers from the recent world recession. Demand growth will be highest in Asia, where apparel is a large application; growth will also occur in niche markets, such as medical tape.

Regulations concerning smoking in Asia, Central and Eastern Europe, Africa, the Middle East, and Central and South America are fewer and not as restrictive as in, for instance, the United States; additionally, social pressure to stop smoking is not as strong in these regions. Smoking prevalence is declining in North America, Western Europe, Japan and Australia. While regulations influence smoking on a regional level, global consumption of cigarettes is forecast to grow at an average annual rate of approximately 0.7% during 2010-2020. This is a slower rate of growth than during 2000-2010, when global cigarette consumption grew at an average annual rate of approximately 1.0%; this higher growth was due largely to increased cigarette consumption in Asia, where population growth and increases in living standards experienced a period of significant growth.

Cellulose acetate cigarette filter tow is not at risk for substitution by competing products. Replacement of cellulose acetate textile fiber primarily by polyester fibers has largely occurred. China will strengthen its position as the largest single consumer of cigarette filter tow and will likely attract the most investment for expansions and new construction of cellulose acetate cigarette filter tow facilities.

The purified cotton linters or cellulose is fed to the acetylator containing acetic anhydride and acetic acid and conc. Sulphuric acid and acetylation is carried out at 25 ~ 30 ℃. The reaction mixture called acid dope is allowed for ripening for about 10~20 hrs. During reopening conversion of acetate groups takes place. After reopening, the mixture is diluted with water with continuous stirring. During the process flakes acetate rayon is precipitated which is dried and send to spinning bath where dry spinning of acetate rayon takes place by dissolving in solvent and passing through spinnerated. The solvent is evaporates by hot air. The dope coming from the spinnerate is passed downwards to feed roller and finally to bobbin where spinning is done at higher speed. Process flow diagram for the manufacture of cellulose acetate is given in Figure 5.11.

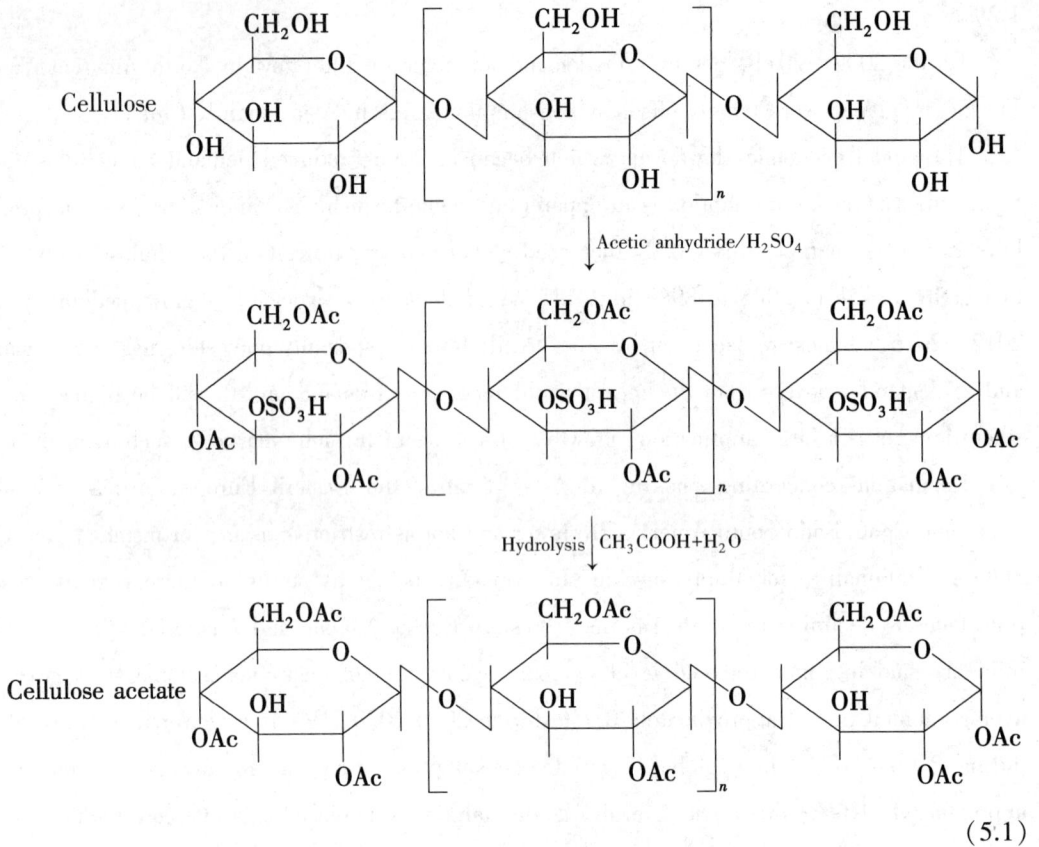

Cellulose

$\downarrow$ Acetic anhydride/H_2SO_4

Hydrolysis $\downarrow$ $CH_3COOH+H_2O$

Cellulose acetate

$$(5.1)$$

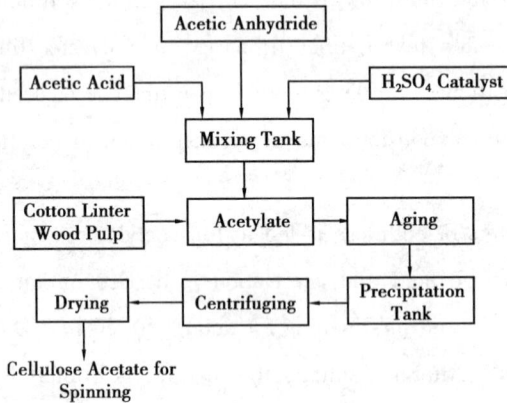

Figure 5.11　Process flow diagram for the manufacture of Cellulose acetate

5.3.4　**Polyester fibers** [10]

The most commercially important aromatic polyester is poly (ethylene terephthalate) (PET). Among the aromatic polyesters, PET is considered as "work horse". PET is otherwise known as polyethylene glycol terephthalate, ethylene terephthalate polymer, poly (oxy-1, 2-ethanediy-loxycarbonyl-1, 4-phenylene dicarbonyl), terephthalic acid-ethylene glycol polyester. It is also known as 2GT. PET is a white or light cream material, has high heat resistance and chemical stability and is resistant to acids, bases, some solvents, oils and fats. The unit molecular weight of PET is 192 and the chemical structure of PET is given below:

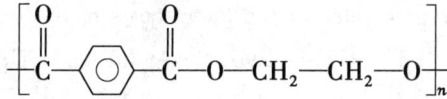

$$\left[\begin{matrix} & O & & O \\ & \| & & \| \\ -C- & \langle\!\bigcirc\!\rangle & -C-O-CH_2-CH_2-O- \end{matrix}\right]_n$$

Polyethylene terephthalate (PET) polymer is produced from ethylene glycol and either dimethyl terephthalate (DMT) or terephthalic acid (TPA). Polyester filament yarn and staple are manufactured either by direct melt spinning of molten PET from the polymerization equipment or by spinning reheated polymer chips. Polyester fiber spinning is done almost exclusively with extruders, which feed the molten polymer under pressure through the spinnerettes. Filament solidification is induced by blowing the filaments with cold air at the top of the spin cell. The filaments are then led down the spin cell through a fiber finishing application, from which they are gathered into tow, hauled off, and coiled into spinning cans. Depending on the desired product, post-spinning operations vary but may include lubrication, drawing, crimping, heat setting, and stapling.

Formation of PET consists of two main reactions namely: ①esterification (or) precondensation and ②poly-condensation. The esterification reaction is conducted in excess of MEG. The first step produces a prepolymer which contains bis (hydroxyethyl terephthalate) (BHET) (or) diethylene glycol terephthalate and short chain oligomers

A.Hopper fed with PET chips
B.Spinning vessel
C.Heated grid
D.Dope(molten PET)
E.Spinning jet (spinneret)
F.Cold air
G.Input feed rollers
H.Deflector
I.Output feed roller

Figure 5.12　Schematic melt spinning process for PET fiber

and by-products namely water (or) methanol depending upon the raw material used namely PTA or DMT respectively. In the case of the PTA route, the operating temperature was $240 \sim 265$ ℃ and pressure was 0.4 MPa during precondensation. In order to expedite the rate of polycondensation, a catalyst such as antimony acetate, antimony trioxide, germanium dioxide or titanium can normally be used. In addition to catalyst, stabiliser namely phosphoric acid, phosphorus acid, trimethyl phosphate, triethyl phosphate should be used to stabilize metal ions such as manganese, zinc, calcium, etc. and deactivate them when used as transesterification catalyst. The most effective catalysts for the transesterification of DMT with MEG were the acetates of zinc, lead (Ⅱ), mercury (Ⅱ) together with cobalt (Ⅲ) acetylacetonate and antimony trioxide. A detailed review for esterification and transesterification reaction using titanium catalyst has been reported elsewhere. Theoretically, under such conditions, all chains should terminate with hydroxyl end groups. However, degradation reaction produces a certain amount of carboxyl end groups, which indicates the extent of degradation that had taken place in the melt.

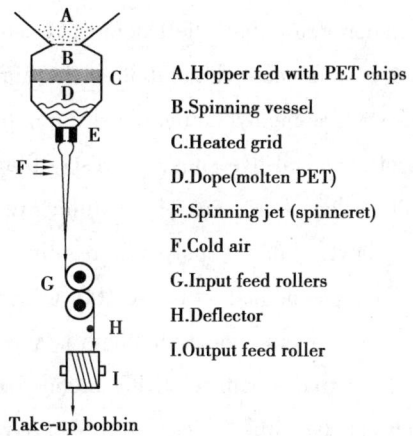

141

In the polycondensation reaction, there are two chain growth reactions. They are: ① polyesterification between chain ends with carboxyl and hydroxyl end groups with elimination of water and ② polytransesterification between hydroxylethyl end groups and ester end groups with the elimination of MEG, which predominates in the later stages of the reaction. Polycondensation was carried out in vacuum (0.13 kPa) at $275 \sim 290$ ℃. At the end of polycondensation, the molten polymer was quenched in water to obtain strands, which were granulated into chips. Intrinsic viscosity (Ⅳ) obtained was in the range of 0.5 to 0.7 dL/g and the residual acetaldehyde (AA) content was approx. 50×10^{-6}.

Polyesters have good resistance to most mineral acids but concentrated sulphuric acid dissolves polyesters with partial decomposition. Polyesters display excellent resistance to conventional bleaching agents, cleaning solvents and surfactants and the degree of crystallinity and molecular orientation determines the extent of resistance. Basic substances attack polyester fibres in two ways. Strong alkalis cause dissolution of the fibre surface. Weak bases such as ammonia and other organic bases such as methyl amine penetrate into the non-crystalline regions of the structure. Because of the lack of chemical dye sites, polyester fibres are usually dyed with disperse dyes such as amacron, artisil, calcosperse, cekryl, celliton, dispersol, duranol, esterophile, foron, genecron, harshaw ester, latyl, palanil, polydye, resolin, samaron, setacyl, terasil, etc. Polyester fibers are dyed from an aqueous bath at above 100 ℃ or by the use of a carrier such as biphenyl, phenyl salicylate. The rate of dyeing polyester fibers is slower than that of cellulose triacetate or acetate.

PET grades with relatively higher molecular weights are used for making industrial filaments, which can be either thick for rubber tyres, conveyor belts, seat belts, hoses and ropes, coated fabrics, etc. or relatively thinner for sewing threads, light-weight coated fabrics, etc. Staple fiber finds major use in making blended fabrics, low denier fiber for blending with cotton and coarser fiber for blending with wool. Fabrics made from PET microfilaments are breathable and water-repellent with soft drape and pleasant feel.

5.3.5 Nylon fiber

Nylon is a generic word representing a class of polymers as polyamides. Nylon was a product of a basic research started by Wallace Corothers at DuPont in 1928. The polymer age really started when the new synthetic nylon fiber was introduced in the market in 1938. Nylon captured public imagination, as four million pairs of nylon stockings were sold in the first few hours of sale on May 15, 1940. During World War Ⅱ, nylon fabrics were used as waterproof tents, lightweight parachutes among other things. In the process an important principle was established that it is possible to chemically link hundreds of simple molecules of a kind or more, to make polymers with unique set of properties. One interesting example is the ability of molten polymer undergoing large deformations, which allows continuous production of fine filaments from thicker extrudates and this property of deformation is in turn connected to cumulative interactions of long chain polymeric molecules [10].

Nylon 66 (or 6,6-Nylon) is a condensation polymer formed by the condensation polymerization of Adipic acid (a dibasic acid) and hexamethylene diamine (a diamine). In the nomenclature, the first number refers to the number of carbon atoms in the diamine and the second number designates the number of carbon atoms in the dibasic acid [11].

Nylon 6 is a polymer of caprolactam. The number refers to the total carbon atoms in the ring, which can vary from 5 to 12.

Both Nylon 6 and Nylon 66 have similar properties and can be used as thin films and fibres; and for extrusion and injection molding in plastics. The major difference in the physical properties of these two nylons is that the melting point of Nylon 66 is 40~45 ℃ higher than Nylon 6. Moreover, Nylon 66 is harder, rigid and its abrasive-resistance is lower than that of Nylon 6.

In the production of Nylon 66(Figure 5.13), first, the amine and acid is mixed in an aqueous solution, with the pH adjusted at 7.8 to form Nylon salt. The salt is then concentrated under vacuum. It is thereafter charged to an autoclave where the rest of the water is removed. The temperature is gradually increased to 280 ℃ in the autoclave to complete the polymerization and the water of condensation is removed.

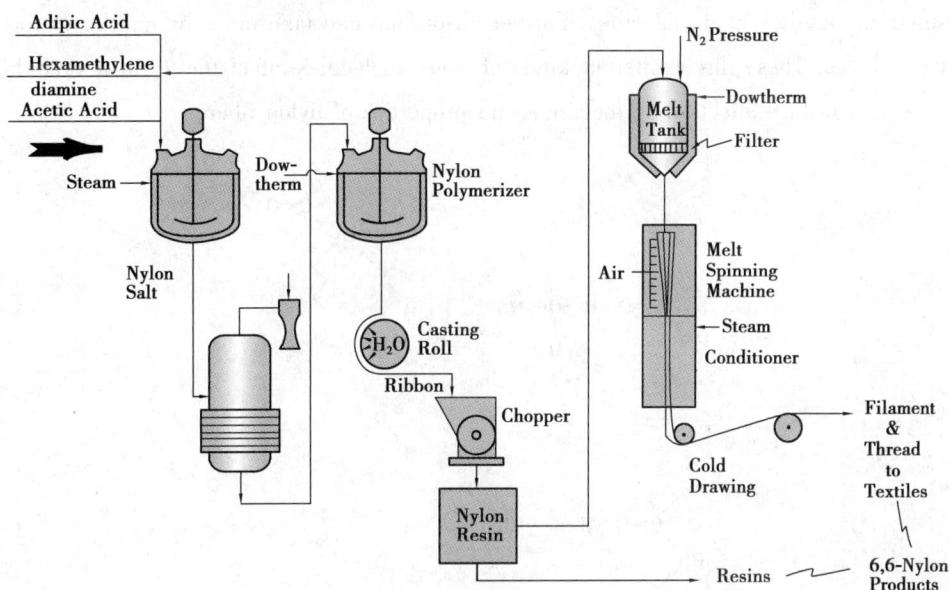

Figure 5.13   Manufacture of Nylon fibers

The final product has a molecular weight of 12 000 ~ 16 000. It is extruded as ribbons onto chilling rolls. Then, it is sent to a chipper which produces small chips, which possesses the convenience of storage and rehandling. These chips are melted, metered through high pressure pumps. After filtration, these are passed via a melt spinnerette to produce nylon fibers. Upon cooling, these filaments harden and are wound on bobbins at a rate of 750 meters per minute or even higher. In order to give the fiber desirable textile properties, the threads are stretched to about

143

4 times its original length.

The production of Nylon 6 is similar to that of Nylon 66, the only difference being the nature of polymerization. Nylon 6 is manufactured by the step-wise condensation of caprolactam with no net water removal. However, to maintain the thermal equilibrium between monomer and polymer at the melting and spinning temperatures is the only difficult task. Monomer retained on the fibers accounts up to 10%, so the fibers are water-washed to remove the soluble caprolactam retained.

In the melt spinning process, essentially dry nylon is melted and spun in filament form. Filaments absorb moisture in the various steps of production process, i.e. during quenching, spin finish application and further processing. Thus, T_g dependence on varying moisture conditions during production processes is quite critical. T_g is known to have a non-linear relationship with moisture regain for amorphous nylons. There have been several attempts to explain the T_g and moisture regain relationship; a reasonable explanation is obtained by three step moisture absorption by nylons. The nature of absorption of molecules is shown in Figure 5.14. In the process of moisture absorption from dry nylon, the initial water molecules form double hydrogen bonds between two carbonyl groups by means of free electron pairs on oxygen atoms and may be assessed as firmly bound. Water molecules are also attached by replacing the hydrogen bond between two carbonyl and amide groups and may be classified as loosely bound molecules. Further absorption can take place by multilayer formation of water molecules. These three different kinds of water molecules affect the T_g in a very different manner and this in turn affects the processing and properties of nylon fibers.

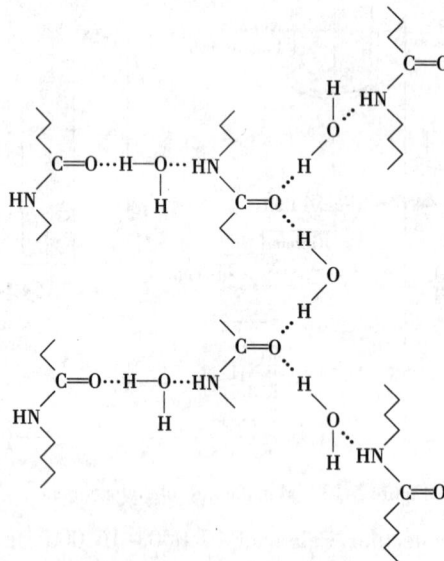

Figure 5.14 Interaction of amide groups with water in nylon

Modulus of nylon fiber at low temperature has shown that initially, as regain increases from the dry state, modulus increases. This implies that the initial water molecules are tightly bound and have a partial specific volume less than unity, thus making the state very rigid. Variation of torsional rigidity of nylon 6 fibers is a function of regains and observes a maximum rate of change at 3% ratio regain. This corresponds to increased mobility of loosely bound water molecules, as observed from NMR measurements at these regain levels. Stress ageing experiments on nylon 6 filaments indicate that stress induced microcrystal formation takes place during moisture absorption, in the regain range of about 2%.

Choice of moisture regain in the quenching zone in nylon filament production process is quite critical, as the moisture regain is changing rapidly. An RH level of 70% will be helpful, as the filaments will quickly attain the regain levels of about 3% in the highly dynamic filament formation process. Drawing process of nylon 6 is also carried out at around 55% RH corresponding to 3% regain levels. Choice of this level is related with minimum changes in T_g of the fiber with fluctuations in relative humidity in this range.

Melting of nylon fibers is typical to semi-crystalline polymers, occurring in a range of temperatures. The melting range of nylon 6 is in the temperature range of 215~228 ℃, while the nylon 66 melting range is between 250 ℃ and 265 ℃. Melting of nylons is dependent on concentration of amide groups and number of CH_2 groups linking these groups. The variation of melting with number of CH_2 groups per monomer unit is shown in Figure 5.15. The figure further indicates that the melting also depends on symmetry of the structural units. Heat of fusion of α-crystal form for nylon 6 is 64 cal/g., whereas nylon 66 has a value of 61 cal/g.

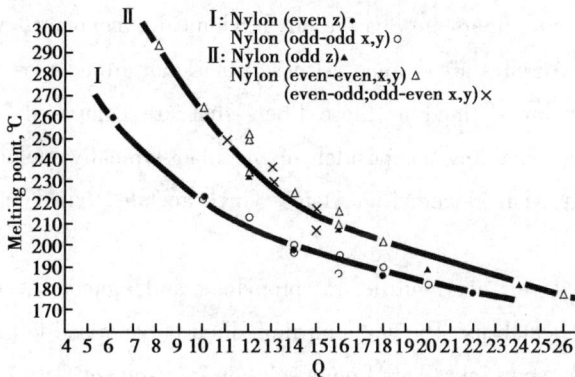

Figure 5.15   Melting point of aliphatic polyamides

A reasonable morphological model for nylon fibers is described in terms of the three phase model with microfibrillar and inter-microfibrillar regions as shown in the Figure 5.16. It should be noted that a two phase model, consisting of crystalline and amorphous regions in fibers, does not adequately describe the fiber structure. The width of the microfibrils ranges between 60 Å and 200 Å. The microfibrils in turn consists of crystalline and amorphous regions in a series mode. Industrial grade filaments typically have crystalline and amorphous region lengths of about 60 Å and about 30 Å respectively. These crystalline and amorphous regions, arranged in a regular sequence along the fiber axis, act as a micro-lattice. In fact the regularity is sufficient to give diffraction maxima in the small-angle of x-ray scattering. Longitudinal dimensions of microfibrils are not well

145

defined as is the case with this model type. Microfibrils are surrounded by inter-microfibrillar regions and consist of highly oriented molecular chains and are thus in some degree of pseudo-order. Microfibrils in turn form an endless oriented network with branching and fusion more common than endings.

In the drawing process, nylons deform through micro fiber slippage forming a large number of interfibrillar tie molecules which appear as a separate phase. Thus the modulus and strength of nylons are dominated by the amount of interfibrillar extended chains and should be considered as a separate phase.

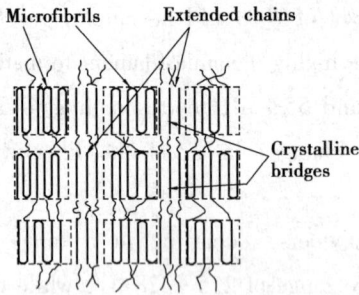

Figure 5.16 Microstructure of highly drawn polyamide fibers

There is a significant interaction between microfibrils in nylons. The extended chains in interfibrillar regions are the strongest and have a profound effect on fiber strength. Thus strength can be increased by shearing off the surface of the microfibrils to make an interfibrillar domain. In the highly oriented state, microfibrils tend to fuse via epitaxial crystallization of extended chain molecules as shown in Figure 5.16.

5.3.6 Acrylic and modacrylic fiber[5]

Acrylic and modacrylic fibers are based on acrylonitrile monomer, which is derived from propylene and ammonia. Acrylics are defined as those fibers that are composed of at least 85 percent acrylonitrile. Modacrylics are defined as those fibers that are composed of between 35 and 85 percent acrylonitrile. The remaining composition of the fiber typically includes at least one of the following: methyl methacrylate, methyl acrylate, vinyl acetate, vinyl chloride, or vinylidene chloride.

The starting materials for acrylonitrile are propylene and ammonia, which are reacted with oxygen in the presence of catalysts. The acrylonitrile is then polymerised to produce polyacrylonitrile (PAN). The PAN is then spun into fibers from a solution in a solvent. Two process routes are used, wet spinning in which the fibers are spun into an aqueous coagulation bath and dry spinning in which the fibers are spun into hot air [12].

For the fabrication of acrylic fiber, the acrylic polymer is firstly dissolved in a suitable solvent, such as dimethylformamide or dimethylacetamide. Additives and delusterants are added, and the solution is usually filtered in plate and frame presses. The solution is then pumped through a manifold to the spinnerettes (usually a bank of 30 to 50 per machine). At this point in the process, either wet or dry spinning may be used to form the acrylic fibers. The spinnerettes are in a spinning bath for wet spun fiber or at the top of an enclosed column for dry spinning. The wet spun filaments are pulled from the bath on takeup wheels, and then washed to remove more solvent. After washing, the filaments are gathered into a tow band, stretched to improve strength, dried, crimped, heat set, and then cut into staple. The dry spun filaments are gathered into a tow band, stretched, dried, crimped, and cut into staple. The modacrylic fibers contain halogen comonomers such as vinyl chloride or vinylidene chloride, and have flame-retardant properties.

Acrylic fibers are soft, flexible and have a high loft. For this reason they are widely used in knitted apparel end-uses such as sweaters and socks. In addition to knitted apparel, home furnishing and blankets are other important applications due to its excellent heat retention.

5.3.7　Polyolefin fiber

Polyolefin fibers are molecularly oriented extrusions of highly crystalline olefinic polymers, predominantly polypropylene. Melt spinning of polypropylene is the method of choice because the high degree of polymerization makes wet spinning or dissolving of the polymer difficult. The fiber spinning and processing procedures are generally the same as described earlier for melt spinning. Polypropylene is also manufactured by the split film process in which it is extruded as a film and then stretched and split into flat filaments, or narrow tapes, that are twisted or wound into a fiber. Some fibers are manufactured as a combination of nylon and polyolefin polymers being melted together in a ratio of about 20% nylon 6 and 80% polyolefin such as polypropylene, and being spun from this melt. Polypropylene is processed more like nylon 6 than nylon 66 because of the lower melting point of 203 ℃ (397 ℉) for nylon 6 versus 263 ℃ (505 ℉) for nylon 66.

Very large amounts of polypropylene homopolymers are used for fiber manufacture in Europe and the United States. These applications exploit the wide range of physical forms, including increasing amounts of versatile nonwoven fabrics. Mono-axial orientation can be applied to conventional spinneret yarns, as with polyamides and polyesters, and to flat tapes made from extruded film. These differ in the following respects (Table 5.2).

Table 5.2　Polypropylene fiber processes

Process	Filament count, tex*	Product
Long spin	0.2 — 3.0	high-tenacity monofilament; drawing integral or separate; high output
BCF yarn	0.2 — 2.0	special case of long spin making only bulked continuous fiber
Spunbonded	0.2 — 2.0	venturi haul off, no 2nd stage draw. bonded mat output
Shortspiri	0.2 — 40	compact unit; tow stretched and cut in line for staple
Melt blown	0.002 — 0.02	low orientation, very fine fiber; only bonded mat output
Fibrillated yarn	110 — 500	oriented slit film; fibrillated for baler twine, rope, etc.
Weaving tape	ca.110	nonfibrillated slit film for carpet backing, sacks, etc.
Strapping tape	500 — 1 000	thick, oriented tape as a steel alternative

* tex = weight in grams of 1 000 m of yarn.

In the spinneret type yarns, here melt is extruded through a die plate perforated by many small holes to generate individual thread lines. In the long spin process, which has integral spinning and finishing stages as well as out of line drawing options, there may be 50~250 holes per spinneret. An air cooling gap of 2~5 in is needed between the die plate and the wind-up roll. Line speeds up to

1 000 m/min at the spinning stage, increasing to 3 000 m/min during solid drawing over hot rolls, call for complex and expensive haul off, drawing, and wind-up sections. Product may be packed off as continuous yarn, tow, or staple. Suitable polymers have MFI (230 ℃/2. 16 kg) in the range 12~25 dg/min, with narrower molecular mass distribution grades offering some advantage with such high rates of melt draw. These are high-throughput plants, best suited to long runs of a single grade of product [13].

During processing, lubricant and finish oils are added to the fiber, and some of these additives are driven off in the form of aerosols during processing. No specific information has been obtained to describe the oil aerosol emissions for polyolefin processing, but certain assumptions may be made to provide reasonably accurate values. Because polyolefins are melt spun similarly to other melt spun fibers (nylon 6, nylon 66, polyester, etc.), a fiber similar to the polyolefins would exhibit similar emissions. Processing temperatures are similar for polyolefins and nylon 6. Thus, aerosol emission values for nylon 6 can be assumed valid for polyolefins.

5.3.8　Polyvinyl alcohol fiber

Vinyon is a copolymer of vinyl chloride (88%) and vinyl acetate (12%). The polymer is dissolved in a ketone (acetone or methyl ethyl ketone) to make a 23 weight percent spinning solution. After filtering, the solution is extruded as filaments into warm air to evaporate the solvent and to allow its recovery and reuse. The spinning process is similar to that of cellulose acetate. After spinning, the filaments are stretched to achieve molecular orientation to impart strength.

5.3.9　Spandex fiber

Spandex is a lightweight, synthetic fiber that is used to make stretchable clothing such as sportswear. It is made up of a long chain polymer called polyurethane, which is produced by reacting a polyester with a diisocyanate. The polymer is converted into a fiber using a dry spinning technique. First produced in the early 1950s, spandex was initially developed as a replacement for rubber. Although the market for spandex remains relatively small compared to other fibers such as cotton or nylon, new applications for spandex are continually being discovered[14].

Spandex is a synthetic polymer. Chemically, it is made up of a long-chain polyglycol combined with a short diisocyanate, and contains at least 85% polyurethane. It is an elastomer, which means it can be stretched to a certain degree and it recoils when released. These fibers are superior to rubber because they are stronger, lighter, and more versatile. In fact, spandex fibers can be stretched to almost 500% of their length.

This unique elastic property of the spandex fibers is a direct result of the material chemical composition. The fibers are made up of numerous polymer strands. These strands are composed of two types of segments: long, amorphous segments and short, rigid segments. In their natural state, the amorphous segments have a random molecular structure. They intermingle and make the fibers

148

soft. Some of the rigid portions of the polymers bond with each other and give the fiber structure. When a force is applied to stretch the fibers, the bonds between the rigid sections are broken, and the amorphous segments straighten out. This makes the amorphous segments longer, thereby increasing the length of the fiber. When the fiber is stretched to its maximum length, the rigid segments again bond with each other. The amorphous segments remain in an elongated state. This makes the fiber stiffer and stronger. After the force is removed, the amorphous segments recoil and the fiber returns to its relaxed state. By using the elastic properties of spandex fibers, scientists can create fabrics that have desirable stretching and strength characteristics.

The primary use for spandex fibers is in fabric. They are useful for a number of reasons. First, they can be stretched repeatedly, and will return almost exactly back to original size and shape. Second, they are lightweight, soft, and smooth. Additionally, they are easily dyed. They are also resilient since they are resistant to abrasion and the deleterious effects of body oils, perspiration, and detergents. They are compatible with other materials, and can be spun with other types of fibers to produce unique fabrics, which have characteristics of both fibers.

Spandex is used in a variety of different clothing types. Since it is lightweight and does not restrict movement, it is most often used in athletic wear. This includes such garments as swimsuits, bicycle pants, and exercise wear. The form-fitting properties of spandex makes it a good for use in undergarments. Hence, it is used in waist bands, support hose, bras, and briefs.

The development of spandex was started during World War II. At this time, chemists took on the challenge of developing synthetic replacements for rubber. Two primary motivating factors prompted their research. First, the war effort required most of the available rubber for building equipment. Second, the price of rubber was unstable and it fluctuated frequently. Developing an alternative to rubber could solve both of these problems.

At first, their goal was to develop a durable elastic strand based on synthetic polymers. In 1940, the first polyurethane elastomers were produced. These polymers produced millable gums, which were an adequate alternative to rubber. Around the same time, scientists at DuPont produced the first nylon polymers. These early nylon polymers were stiff and rigid, so efforts were begun to make them more elastic. When scientists found that other polyurethanes could be made into fine threads, they decided that these materials might be useful in making more stretchable nylons or in making lightweight garments.

The first spandex fibers were produced on an experimental level by one of the early pioneers in polymer chemistry, Farbenfabriken Bayer. He earned a German patent for his synthesis in 1952. The final developments of the fibers were worked out independently by scientists at DuPont and the U.S. Rubber Company. DuPont used the brand name Lycra and began full scale manufacture in 1962. They are currently the world leader in the production of spandex fibers.

A variety of raw materials are used to produce stretchable spandex fibers. This includes prepolymers which produce the backbone of the fiber, stabilizers which protect the integrity of the polymer, and colorants.

Two types of prepolymers are reacted to produce the spandex fiber polymer back-bone. One is a flexible macroglycol while the other is a stiff diisocyanate. The macro-glycol can be a polyester, polyether, polycarbonate, polycaprolactone or some combination of these. These are long chain polymers, which have hydroxyl groups (—OH) on both ends. The important feature of these molecules is that they are long and flexible. This part of the spandex fiber is responsible for its stretching characteristic. The other prepolymer used to produce spandex is a polymeric diisocyanate. This is a shorter chain polymer, which has an isocyanate (—NCO) group on both ends. The principal characteristic of this molecule is its rigidity. In the fiber, this molecule provides strength.

In days before spandex, how did the corset contour the body effectively? In the eighteenth century, thick quilting and stout seams on the corset shaped the body when the garment was tightly laced. In the early nineteenth century, baleen, a bony but bendable substance from the mouth of the baleen whale, was sewn into seams of the corset (hence the term whalebone corsets), however the late 1 800 s corsets like this were stiffened with small, thin strips of steel covered with fabric. Such steel-clad corsets did not permit movement or comfort. By World War I, American women began separating parts of the corset into two garments—the girdle (waist and hip shaper) and bandeau (softer band used to support and shape the breasts).

Spandex fibers are produced in four different ways including melt extrusion, reaction spinning, solution dry spinning, and solution wet spinning. Each of these methods involves the initial step of reacting monomers to produce a prepolymer. Then the prepolymer is reacted further, in various ways, and drawn out to produce a long fiber. Since solution dry spinning is used to produce over 90% of the world's spandex fibers, it is described.

Essentially all air that enters the spinning room is drawn into the hooding that surrounds the process equipment and then leads to a carbon adsorption system (see Figure 5.17). The oven is also vented to the carbon adsorber. The gas streams from the spinning room and oven are combined and cooled in a heat exchanger before they enter the activated carbon bed.

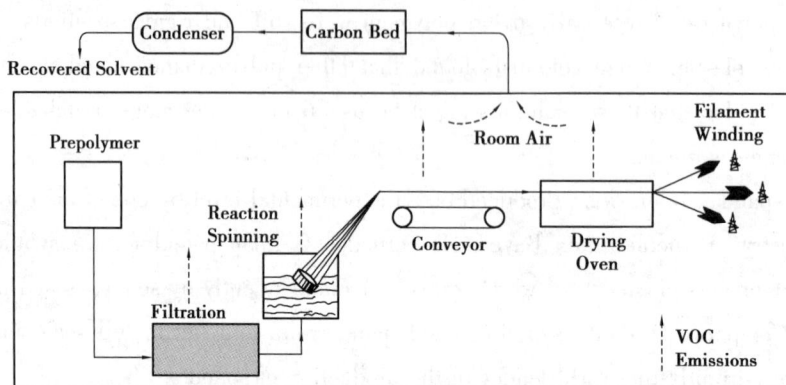

Figure 5.17 Spandex reaction spinning.

In addition to their remarkable stretch and recovery properties, elastanes (spandex) resist perspiration and cosmetic oils, are easily washable, are dyeable and have moderate abrasion resistance. Elastane yarns are often covered with another fiber. This provides more bulk and improves abrasion resistance. The main end-uses for the yarns are garments and other products, where comfort and/or fit are important.

Typical examples are sports and leisure wear, swimming wear, elastic corset fabrics and stockings.

Collection of Exercises

1. List two important characteristics for polymers that are to be used in fiber applications.
2. How do the thermal and mechanical properties of Nomex and Kevlar compare and how are these differences related to their chemical structure? Which polymer would be expected to have the largest characteristic ratio? How can molecular weight of these polymers be determined?
3. What is the minimum length to diameter ratio for a substance to be classified as a fiber?
4. Which is stronger: hydrogen bonding or dipole-dipole interactions?
5. Besides hydrogen bonding, what else is characteristic of fiber molecules?
6. Which is more hydrophobic: cellulose or nylon 6,6?
7. Is rayon produced by wet or dry spinning?
8. Name two fibers that are produced by melt spinning.
9. Which has the lower denier value: a nylon fishline or nylon fiber used to manufacture hosiery?
10. How do the polyester fibers produced originally by Carothers differ from today's polyester fibers, such as Dacron?
11. 天然纤维和化学纤维的种类有哪些?
12. 纤维的分子结构对纤维的性能有何影响?
13. 粘胶纤维为什么被称为再生纤维? 其主要成分是什么?
14. 在纤维生产过程中为什么要进行拉伸? 在拉伸过程中纤维的结构会发生什么变化?
15. 简述合成纤维生产过程中拉伸和热定型的作用。
16. 溶液纺丝和熔体纺丝方法有什么不同? 它们分别适用于什么聚合物?
17. 简述涤纶纤维的优缺点和主要的应用领域。
18. 写出尼龙66和Kevlar纤维的分子结构,讨论其结构与性能的关系。
19. 说明腈纶纤维中加入第二、三弹体对纤维的性能有哪些改进。
20. 再生纤维有哪些新品种? 从可持续发展的角度来看,还有哪些资源可以用于生产纤维?

REFERENCES

［1］What is textile fiber? http://textilefashionstudy.com/，网站资料.

［2］What is Cotton? http://www.ifc.net.au/，网站资料.

［3］http://www.engr.utk.edu/mse/pages/Textiles/Cotton%20fibers.htm，网站资料.

［4］http://www.sewing.org/，网站资料.

［5］Compilation of air pollutant emission factors Vol I：Stationary Point and area sources（Section 6.9 synthetic fibers），US Environmental Protection Agency，1995.

［6］http://nptel.ac.in/courses/103103029/module7/lec40/2.html.

［7］http://www.yarnsandfibers.com/.

［8］http://www.fibersource.com/fiber.html.

［9］https://www.ihs.com/products/cellulose-acetate-and-triacetate-chemical-economics-handbook.html.

［10］B. L. Deopura，R. Alagirusamy，M. Joshi，B. Gupta，Polyesters and polyamides，CRC Press，Boca Raton，USA，2008.

［11］http://nptel.ac.in/courses/103103029/module7/lec40/3.html.

［12］http://www.cirfs.org/manmadefibres/fibrerange/Acrylic.aspx.

［13］Markus Gahleitner，Christian Paulik，Ullmann's Encyclopedia of Industrial Chemistry（Polypropylene），Wiley，2011.

［14］http://www.madehow.com/Volume-4/Spandex.html，Made How，网站资料.

CHAPTER **6**

RUBBERS

6.1 Introduction [1,2]

Rubber was known to the indigenous peoples of the Americas long before the arrival of European explorers. In 1525, Padre d'Anghieria reported that he had seen Mexican tribespeople playing with elastic balls. The first scientific study of rubber was undertaken by Charles de la Condamine, when he encountered it during his trip to Peru in 1735. A French engineer that Condamine met in Guiana, Fresnau studied rubber on its home ground, reaching the conclusion that this was nothing more than a "type of condensed resinous oil".

The first use for rubber was an eraser. It was Magellan, a descendent of the famous Portuguese navigator, who suggested this use. In England, Priestley popularized it to the extent that it became known as India Rubber. The word for "rubber" in Portuguese—borracha—originated from one of the first applications for this product, when it was used to make jars replacing the leather borrachas that the Portuguese used to ship wine.

Returning to the works of Condamine, Macquer suggested that rubber could be used to produce flexible tubes. Since then, countless craftsmen have become involved with rubber: goldsmith Bernard, herbalist Winch, Grossart, Landolles and others. In 1820, British industrialist Nadier produced rubber threads and attempted to use them in clothing accessories. This was the time when America was seized by rubber fever, and the waterproof footwear used by the indigenous peoples became a success. Waterproof fabrics and snow-boots were produced in New England.

In 1832, the Rosburg factory was set up. Unfortunately, cold weather affected goods made from non-vulcanized natural rubber, leaving them brittle and with a tendency to gum together if left in the sun, all discouraging consumers. After a long period attempting to develop a process to upgrade rubber qualities (such as including nitric acid) that almost ruined him, in 1840 Goodyear

153

discovered vulcanization, quite by accident.

An interesting fact: in 1815, a humble sawyer—Hancock—became one of the leading manufacturers in the UK. He had invented a rubber mattress and through an association with MacIntosh he produced the famous waterproof coat known as the "macintosh". Furthermore, he discovered how to cut, roll and press rubber on an industrial scale. He also noted the importance of heat during the pressing process, and built a machine for this purpose.

MacIntosh discovered the use of benzene as a solvent, while Hancock discovered that prior chipping and heating were required in order to ensure that the rubber dissolved completely. Hancock also discovered how to manufacture elastic balls. Finally, in 1842, Hancock came into possession of vulcanized rubber produced by Goodyear, seeking and finding the secret of vulcanization that brought him a vast fortune.

In 1845, R.W. Thomson invented the pneumatic tire, the inner tube and even the textured tread. In 1850 rubber toys were being made, as well as solid and hollow balls for golf and tennis. The invention of the velocípede by Michaux in 1869 led to the invention of solid rubber, followed by hollow rubber and finally the reinvention of the tire, because Thomson's invention had been forgotten. The physical properties of rubber were studied by Payen, as well as Graham, Wiesner and Gérard.

Finally, Bouchardt discovered how to polymerize isoprene between 1879 and 1882, obtaining products with properties similar to rubber. The first bicycle tire dates back to 1830, and in 1895 Michelin had the daring idea of adapting the tire to the automobile. Since then, rubber has held an outstanding position on the global market.

As rubber is an important raw material that plays a leading role in modern civilization, chemists soon became curious to learn more about its composition in order to synthesize it. In the 19 century, work focused on this objective, soon discovering that rubber is an isoprene polymer.

The Russians and the Germans broke fresh ground in their efforts to synthesize rubber. But the resulting products were unable to compete with natural rubber. It was only during World War I that Germany pressured by circumstances had to develop the industrialized version of this synthetic product. This was the springboard for the massive development of the synthetic rubber industry all over the world, producing elastomers (Figure 6.1).

Figure 6.1　Elastomer moulding after World War Ⅱ at James Walker

6.2   The nature of rubbers

The molecular structure of an rubber consists of random coils connected by crosslinks. Irregularity is essential to prevent crystallinity. Irregularity can be due to geometric isomers where the cis configuration gives best rubber properties even though the double bond restrains motion of the bonded carbon atoms. The trans configuration contributes to a regular planar zig-zag conformation that is crystallisable. A fully saturated hydrocarbon polymer can be elastomeric if substituents are in atactic configuration, or if it is a random copolymer where the segments cannot co-crystallise.

A statistical description of polymer elasticity is based upon a random distribution of chain links forming a random coil conformation of the macromolecules. A random coil is more thermodynamically stable compared with the fully extended chain because there are an infinite number of random coils, while there is only one fully extended chain. Random coil statistics is self-avoiding in that the model considers the excluded volume of the molecule. The assumption of a random coil conformation allows prediction of structural characteristics including end-to-end distance, radius of gyration, contour length, persistence length, characteristic ratio [3].

Rubbers are mainly synthesised using chain growth polymerisation because high molar mass is a feature of this mechanism and high molar mass is required for high chain extension. Rubbers synthesised by the mechanism with radical initiation. The hydrocarbon types are such as polybutadiene, poly(butadiene-*co*-styrene), poly(butadiene-*co*-acrylonitrile) and the fluorocarbon rubbers. Polyisobutylene is initiated using cations. Block copolymers poly(styrene-*b*-butadiene-*b*-styrene) is initiated by anions. Poly(ethylene-*co*-propylene) and its diene terpolymers are synthesis by initiation with co-ordination metal catalysts.

Elastomers are based on polymers which have the property of elasticity. They are made up of long chains of atoms, mainly carbon, hydrogen and oxygen, which have a degree of cross-linking with their neighbouring chains. It is these cross-linking bonds that pull the elastomer back into shape when the deforming force is removed [4].

The chains can typically consist of 300 000 or more monomer units. They can be composed of repeated units of the same monomer, or made up of two or more different monomers. Polymers made up of two types of monomer are known as copolymers or dipolymers, while those made from three are called terpolymers (Figure 6.2–Figure 6.4).

Elastomers are arguably the most versatile of engineering materials. They behave very differently from plastics and metals, particularly in the way they deform and recover under load.

They are complex materials that exhibit unique combinations of useful properties, the most important being elasticity and resilience. All elastomers have the ability to deform substantially by stretching, compression or torsion and then return almost to their original shape after removal of the

155

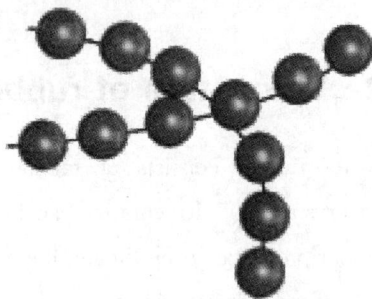

Figure 6.2　Single monomer units polymerised to form a polymer

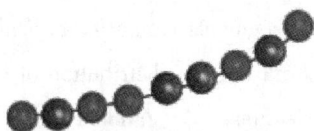

Figure 6.3　Two different monomers
form a copolymer (or dipolymer)

Figure 6.4　Three different monomers
form a terpolymer

force causing the deformation.

Their resilience enables them to return quickly to their original shape, enabling for example dynamic seals to follow variations in the sealing surface.

Elasticity is the ability of a material to return to its original shape and size after being stretched, compressed, twisted or bent. Elastic deformation (change of shape or size) lasts only as long as a deforming force is applied, and disappears once the force is removed (Figure 6.5).

Figure 6.5　Elastomer sample undergoing tensile testing

The elasticity of elastomers arises from the ability of their long polymer chains to reconfigure themselves under an applied stress. The cross-linkages between the chains ensure that the elastomer returns to its original configuration when the stress is removed. As a result of this extreme flexibility, elastomers can reversibly extend by approximately 200% ~ 1 000%, depending on the specific material. Without the cross-linkages or with short, uneasily reconfigured chains, the applied stress would result in a permanent deformation.

Resilience as applied to elastomers is essentially their ability to return quickly to their original shape after temporary deflection. In other words, it indicates the speed of recovery, unlike compression set, which indicates the degree of recovery.

When an elastomer is deformed, an energy input is involved, part of which is not returned when it regains its original shape. That part of the energy which is not returned is dissipated as heat in the elastomer. The ratio of energy returned to energy applied to produce the deformation is defined as the material's resilience.

Most elastomers possess a number of other useful properties, such as:

- Low permeability to air, gases, water and steam
- Good electrical and thermal insulation
- Good mechanical properties
- The ability to adhere to various fibers, metals and rigid plastics.

Also, by proper selection of compounding ingredients, products with improved or specific properties can be designed to meet a wide variety of service conditions.

This remarkable combination of properties is the reason elastomers serve a vast number of engineering needs in fields dealing with sealing, shock absorbing, vibration damping, and electrical and thermal insulation.

Most types of elastomers are thermosets, which gain most of their strength after vulcanisation—an irreversible cross-linking of their polymer chains that occurs when the compound is subjected to pressure and heat. Thermoplastic elastomers, on the other hand, have weaker cross-linking and can be moulded, extruded and reused like plastic materials, while still having the typical elastic properties of elastomers(see Figure 6.6).

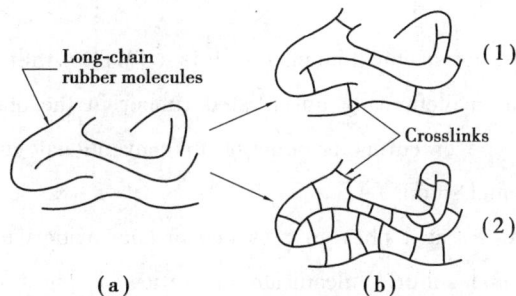

Figure 6.6   Effect of vulcanization on rubber molecules:
(a) raw rubber, and (b) vulcanized (cross-linked) rubber.
Variations of (b) include:(1) soft rubber, low degree of
cross-linking; and (2) hard rubber, high degree of cross-linking

6.3 Rubbers compounding and ingredient

The different compounding ingredients used in rubber latex can be grouped into curing agents, sulfur, accelerators, antioxidants, fillers, pigments, stabilisers, thickening and wetting agents, and other ingredients such as: heat sensitisers, plasticisers, viscosity modifiers, and so on [5].

(1) Curing agent: sulfur

Sulfur is the universal vulcanising agent for natural rubber and also for synthetic rubbers, which contain olefinic unsaturation in the polymer chain, whether these polymers are in latex form or in the form of dry rubber. Sulfur is the main vulcanising agent for natural rubber, synthetic polyisoprene, styrene-butadiene rubber, acrylonitrile-butadiene rubber, polybutadiene rubber, and so on. The crosslinks formed during sulfur vulcanisation of olefinically unsaturated rubber are of three types: monosulfidic, disulfidic and polysulfidic. The relative properties of above crosslinks have an implication in the mechanical and ageing behaviour of vulcanisates. Monosulfidic and disulfidic crosslinks give better ageing resistance compared to polysulfidic linkage, whereas the initial tensile properties are better for a rubber vulcanisates with polysulfidic linkage (see Figure 6.7). When the amount of sulfur used is high, a higher percentage of polysulfidic linkage is formed.

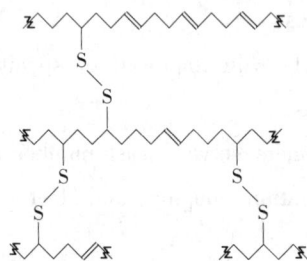

Figure 6.7   Sulfur forms bridges between hydrocarbon chains (cross-links)

Sulfur to be used for latex compound should be of good quality and easily dispersed in water. Colloidal sulfur is preferred for latex compounds, which is obtained by a reaction between hydrogen sulfide and sulfur dioxide in an aqueous medium.

Thiurams, for example, tetramethylthiuram disulfide (TMTD) with disulfidic linkage can be used as a vulcanising agent in olefinically unsaturated rubber in the absence of elemental sulfur (sulfurless curing). This type of curing is superior to conventional curing for heat resistance, oxidative aging resistance, and so on.

Butyl xanthogen disulfide (at 4 phr) in presence of zinc oxide can be used for vulcanising rubber latex without elemental sulfur. Vulcanisate properties of this system are inferior to those obtained using the thiuram system.

(2) Accelerators

The rate of sulfur vulcanisation can be increased by the addition of accelerators. The most important class of accelerators used in latex industry are metallic and amine dialkyl dithiocarbamate, thiazoles and thiurams function as secondary accelerators.

Dithiocarbamates are a class of accelerators used as primary accelerators in latex compounds. It can be in the form of alkali metal salts such as sodium diethyl dithiocarbamate (SDC) or zinc salts such as zinc dimethyl dithiocarbamate (ZMDC). An important difference between the ammonia and alkali metal salts compared to the polyvalent metallic ions is that the former are soluble in water, whereas the latter are not. Water insoluble solids are incorporated in latex as dispersions in water. Table 6.1 shows the preparation of a sulfur dispersion.

Table 6.1 Preparation of sulfur * **dispersion** (50%)

Ingredient	Parts by weight
Sulfur	100
Dispersol F Conc	4.0
Distilled water	96.0

* Ball milled for 72 hours

Commonly used dialkyl dithiocarbamates in latex compounds are zinc diethyl dithiocarbamate (ZEDC), SDC, and piperidinium pentamethylene dithiocarbamate.

The accelerating activity of various dithiocarbamates differs considerably. ZDEC is of intermediate activity and it tends to cause gradual thickening in ammonia preserved natural rubber latex (NRL) under normal storage conditions due to the slow liberation of zinc ions. An exception to this rule is zinc pentamethylene dithiocarbamate. Latex films turn brown in the presence of dithiocarbamate and copper due to the formation of copper dithiocarbamate.

Xanthates are very reactive accelerators. They are active even at room temperature. They are somewhat unstable and are invariably accompanied by a bad odour. This may be due to a small amount of carbon disulfide, which is evolved during their decomposition on storage. Alkali metal xanthates are water soluble whereas heavy metal salts are insoluble. Typical examples are zinc isopropyl xanthate, sodium isopropyl xanthate, zinc-*n*-butyl xanthate.

Thiazoles are used as secondary accelerators in combination with dithiocarbamates. They impart lower compression set and higher modulus and load bearing capacity. Two thiazoles which are most commonly used in latex compounding are sodium mercaptobenzothiazole (SMBT), and zinc mercaptobenzothiazole (ZMBT). SMBT are usually prepared by dissolving mercaptobenzothiazoles in a slight excess of sodium hydroxide solution. The most suitable thiazole accelerator for latex work is the water insoluble ZMBT. This may be prepared by a reaction between sodium mercaptobenzothiazole solution and zinc sulfate solution. ZMBT can be used in place of ZDEC to get desirable technological properties such as high modulus, and so on.

Thiurams are used as a secondary accelerator along with dithiocarbamates. Some typical examples are tetramethylthiuram monosulfide (TMTM), tetraethylthiuram disulfide, dipentamethylenethiuram disulfide (DPTD), dipentamethylenethiuram tetra sulfide. All these accelerators are insoluble in water and so they are incorporated in latex as dispersions. The cure with these accelerators is comparatively slow but their activity can be improved by incorporating sulfur bearing compounds such as thiourea in the compound.

(3) Antioxidants

The ageing characteristic of rubber latex vulcanisates is better compared to dry rubber, since it is not subjected to any degradation during processing (in latex processing there is no mastication or exposure to high temperatures). For NRL products the aging resistance is further improved by the presence of naturally occurring rubber constituents which function as antioxidants. Similarly some of the vulcanisation chemicals such as ZDEC/zinc mercaptoimidazole, and so on, also improves aging resistance.

Two types of antioxidants are used in rubber compounding: the amine type and the phenolic type.

Amine type antioxidants cause discoloration/staining on ageing and because they are resinous in nature it is difficult to disperse them in the rubber latex.

Phenolic antioxidants are the most commonly used in latex, compounding examples are styrenated phenol (SP), substituted cresols, and so on. Water insoluble liquid antioxidants are incorporated in to the latex as an emulsion in water. The emulsified antioxidant droplets are adsorbed on to the rubber particles as the compound matures. Even if this does not occur they will be expected to disperse rapidly in the rubber phase when the latex is dried down to form a solid deposit.

(4) Fillers and pigments

Inorganic fillers and pigments are added to the latex in order to make it less expensive and to stiffen the product or to colour it. These fillers don't have any reinforcing effect when they are added to latex as they do in dry rubber. If the compounded latex with carbon black as filler is subjected to irradiation by high-energy radiation this causes some reinforcement.

The next sections discuss the important inorganic fillers used in the latex compounding.

1) Kaolinite clays

Kaolinite clays are a class of inorganic fillers, which are commonly incorporated in to latex compounds. The kaolinite clays form an important group of fillers, which are an inexpensive material of fine particle size and they are readily dispersed in water with the aid of small amounts of dispersing agents. Some grades of kaolinite clays can be added directly in dry form. The pH of an aqueous clay slurry is usually in the range of 7 to 8. In some cases the pH is lower and in range of 4.0 to 4.5. The acidity may be readily corrected by the addition of a small amount of potassium hydroxide (KOH).

Kaolinite clay is sometimes added to NRL at a level of 400 phr. At these levels the products are very hard and show virtually no rubbery properties.

2) Calcium carbonate

Naturally occurring forms of calcium carbonate are whiting, chalk, limestone, and so on. All of these are very inexpensive and give poor quality products with a marked tendency to discoloration. Precipitated calcium carbonate is widely used in latex compounds. It may contain small quantities of water-soluble calcium salts, which tend to reduce the stability of the latex. One method of detecting the presence of soluble calcium salts is by the addition of a small amount of sodium carbonate prior to the addition to the latex.

3) Titanium dioxide

The most effective white pigment used in latex compounding is rutile (titanium dioxide). For regular application five parts (phr) is used in latex paints.

4) Lithopone

Lithopone is a mixture of barium sulfate and zinc sulfide, and it may be used as filler. It is used as an inexpensive alternative for titanium dioxide.

5) Barytes

Barytes is precipitated barium sulfate, which has been used with NRL to give filled compounds with good extensibility and elongation at break. The main disadvantage of this pigment is its tendency to sediment rapidly. This is because of its high specific gravity.

6) Carbon black

Carbon black is used as black pigment in latex compounding. The carbon blacks are added to latex in the form of dispersion or slurries after adjustment of the pH to alkaline. Wet ground mica is also used as a filler in latex compounding.

(5) **Stabilisers**

1) Surfactants

These are substances which lower the surface free energy against air and aqueous media, along with interfacial free energy against immiscible organic liquids. One method to classify these agents is based on function:

①Wetting agent

②Dispersing agent

③Dispersion stabilisers

④Emulsifiers

⑤Foam promoters and foam stabilisers

The disadvantage of this classification is that there exists a considerable degree of overlap among different categories. For example, potassium oleate is classified as a foam promoter for latex foam and as a stabiliser for synthetic and NRL. Most surfactants are tolerably efficient in the majority of functions but may be outstandingly efficient in just one respect.

Chemically surfactants are classified as amphoteric, anionic, cationic, and non-ionic, and so on, depending upon the active entity present. Anionic surfactants: In this case the surface activity is attributed to the anion—examples are carboxylates, sulfates, sulfonates, and so on.

Unsaturated straight chain aliphatic carboxylates derived from oleic acid find application as colloidal electrolytes in emulsion polymerisation. The oleates are also used as an emulsifier in water immiscible oils, and as a foam promoter in the manufacture of latex foams.

2) Fugitive soaps

These are ammonium soaps. They lose free amines by vulcanisation. Ammonia is the most volatile base but is rather too volatile for some applications, alternatives include morpholine and triethanolamine. Rosin acid soaps also find applications as emulsifiers, colloidal electrolytes and foam promoters.

3) Sulfonates

These are much more sensitive to acids and heavy metal ions than are the carboxylates. They mainly function as wetting agents and examples of this group are sodium diisopropyl naphthalene sulfonate, sodium dibutyl naphthalene sulfonate, and so on. A well-known compound sodium naphthalene formaldehyde sulfoxylate is prepared by reacting two molecules of sodium naphthenate sulfonate with formaldehyde. This substance is a deflocculating and dispersing agent and it finds application in preparing dispersions of insoluble powder. Sodium salts or esters of sulfonic acid are another group and they find application as wetting and dispersion stabilisers.

4) Sulfate

In general, substances of this class are all strongly surface active and find application as wetting agents and dispersion stabilisers. The typical examples of straight chain alkyl sulfonates include sodium dodecyl sulfate, sodium hexadecyl sulfate and a mixture of them.

(6) **Thickening and wetting agents**

1) Thickening agent

It is frequently necessary to increase the viscosity of latex compounds. Thus, dipping mixes may require to be thickened so that thicker deposits of rubber are obtained or spreading mixes are thickened to prevent the latex from striking through the fabric.

Latex compounds may be thickened in two ways: ① by filling the mix or ② by adding thickening agents. The tolerance of latex for these fillers is limited and their addition may produce undesirable effects in the rubber. It may prove necessary, therefore to add thickening agents, among which a wide range of natural products are available, e.g., gums, casein, glue and gelatine. These are all somewhat unpredictable in effect, are subject to bacterial attack and although they may cause high initial increase in viscosity, this effect decreases on prolonged storage. Furthermore, they have marked effects on the 'handle' of the rubber article and on its resistance to water.

Commonly used thickening agents are sodium carboxymethyl cellulose, polyvinyl alcohol, and so on.

2) Wetting agent

Sometimes the addition of a wetting agent to the latex mix is necessary for successful impregnation of fabrics or fibres with latex. Though a medium speed wetting agent, Calsolene oil HS has been found to assist in obtaining a complete penetration between textile fibres without any danger of destabilising the latex.

3) Calsolene oil HS

Calsolene oil HS, a highly sulfonated oil, is available as a clear, amber coloured liquid, readily soluble in water. Unlike some high speed wetting agents, Calsolene oil HS does not give rise to viscosity changes of the latex compounds.

(7) Other compounding ingredients

1) High styrene resin

High styrene resin latex can be used as a reinforcing resin. High styrene-butadiene co-polymer lattices enhance the stiffness and hardness of the deposits.

There is a progressive increase of modulus and a progressive reduction in elongation at break as the proportion of resin in the vulcanisate is increased. Tensile strength increases at first as the level of resin is increased/and then passes through an optimum and falls off, which may be due to the breakdown of resin particles and the rubber matrix interface. High styrene resin lattices may be used in conjunction with lattices of polychloroprene/acrylonitrile-butadiene and copolymers of styrene-butadiene and natural rubber.

2) Resorcinol formaldehyde resin

The direct condensation of resorcinol and formaldehyde in natural rubber is not easy to effect. Formaldehyde tends to destabilise the latex. Excess ammonia is added to prevent coagulation. Formaldehyde reacts with ammonia and prevents the resin formation. If excess of a fixed alkali is added to prevent coagulation, difficulty may arise because of the faster rate of resorcinol formaldehyde resin formation.

(8) Compounding considerations

When creating a new compound there are three main criteria compounders use to guide them. Listed in order of importance they are as follows:

①Customer requirements

②"Processability"

③Cost

Nearly all new compounds are modifications of some existing formulations. Nowadays, development of a completely new compound is seldom attempted. Moreover, such an attempt is usually unnecessary. In order to be efficient and effective in rubber compounding, chemist should take full advantage of technical information readily available inside as well as outside of his organization. He must be analytical, resourceful, and innovative. The following is a useful procedure to guide compound development.

①Set specific objectives (properties, price, etc.).

②Select base elastomer(s).

③Study test data of existing compounds.

④Survey compound formulations and properties data presented by material suppliers in their literature.

⑤Choose a starting formulation.

⑥Develop compounds in laboratory to meet objectives.

⑦Estimate cost of compound selected for further evaluation.

⑧Evaluate processability of compound in the factory.

⑨Use compound to make a product sample.

⑩Test product sample against performance specification.

Rubber compounding is one of, if not the most difficult and complex subjects to master in the field of rubber technology. Compounding is not really a science. It is part art, part science. In compounding, one must cope with literally hundreds of variables in material and equipment. There is no infallible mathematical formulation to help the compounder. That is why compounding is so difficult a task. Example formula is shown in Table 6.2 and Table 6.3:

Table 6.2   An unfilled natural rubber formulation

Ingredient	Amount(phr*)
Natural rubber	100
Process oil	2
Stearic acid	2
Zinc oxide	5
Antioxidant: 6PPD	1
Sulfur	2.75
Cure accelerator: benzothiazyl disulfide	1
Cure accelerator: tetramethyl thiuram disulfide	0.1

* Parts by weight per 100 parts by weight of rubber

Table 6.3   Carbon black-filled natural rubber formulations for general-purpose engineering use

Ingredient	Amount (phr*)
Natural rubber	100
Process oil	5
Stearic acid	2
Zinc oxide	5
N-550 carbon black	25, 50, 75
Phenylamine antioxidant	1.5
Sulfur	2.5
Cure accelerator: benzothiazyl disulfide	1.0
Cure accelerator: tetramethyl thiuram disulfide	0.1

* Parts by weight per 100 parts by weight of rubber

6.4   Manufacturing techniques for rubbers

6.4.1   Compounding design [4]

Elastomer compounds can be designed for specific purposes by modifying their characteristics through varying the quantities of their constituents. This can range from compounding using diluent fillers and basic ingredients to keep costs down, through to the use of specific additives to produce properties such as high tensile strength or wear resistance.

Varying quantities and the selection of ingredients can heavily influence the end properties of the compound, as illustrated in the following examples.

Figure 6.8 shows the effect on price and performance of varying the concentration of reinforcing and diluent fillers. The reduction in cost by increasing the levels of diluent filler content when compared to the reinforcing filler content needs to be balanced against the lower performance.

The effect of varying the acrylonitrile (ACN) content in a nitrile elastomer is shown in Figure 6.9. Increasing the concentration of ACN can be seen to improve oil resistance, while decreasing its concentration improves low temperature flexibility. This is due to the influence of the ACN as a plastic modifying the rubber influence of the butadiene.

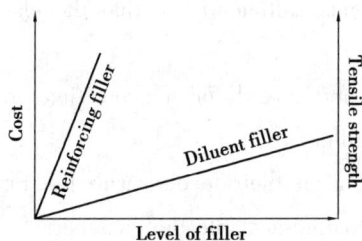

Figure 6.8   Effects of varying concentration of reinforcing and diluent fillers

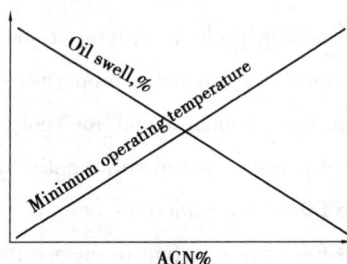

Figure 6.9   Effects of varying the ACN content in a nitrile elastomer

6.4.2   Mixing

Three types of processes are used for mixing the compound ingredients.

(1) Open mill

Here the rubber is banded around the front roll and the ingredients incorporated in the nip (see Figure 6.10).

(2) Internal mixer

The internal mixer has the advantage of being totally enclosed. It mixes a batch of material in

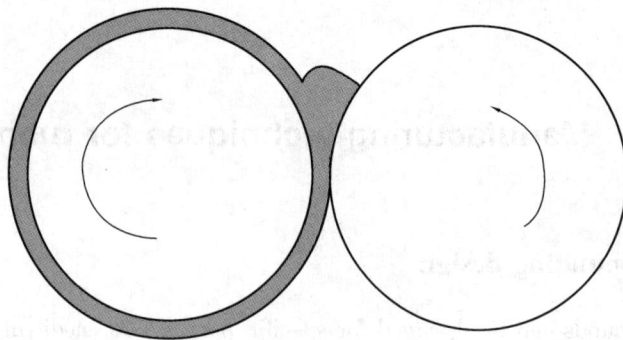

Figure 6.10   Two-roll mill

about 4 to 6 minutes as opposed to up to 30 minutes on an
open mill. In most cases the compound exits the mixer onto a
two-roll mill where it is cooled and compressed into sheet form
ready to be supplied to the manufacturing process (see Figure
6.11).

(3) Continuous mixer

This machine is similar to a long screw extruder, with the
ingredients added via hoppers along the barrel. It is mainly
used where only a few ingredients are added.

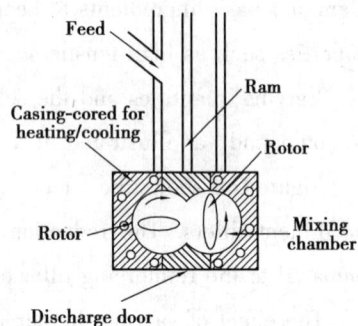

Figure 6.11   Internal mixer

6.4.3   Mixing process

The mixing cycle is crucial in dispersing the ingredients sufficiently so that the elastomer's
physical and fluid resistance properties can be optimised.

In a conventional cycle the polymer is added first and mixed for a short time to ensure
homogeneity and to soften sufficiently to accommodate the fillers.

The fillers are added in one or more stages depending on their levels, with the ram being
lowered after each addition to ensure the material is fully compressed into the chamber.

These days most fillers are automatically weighed and fed directly into the mixer for accuracy
and to avoid contamination. These records are automatically stored in the mixing computer system. If
plasticisers are used they are usually added with the fillers to aid dispersion.

It is important to optimise the chamber volume fill so that the shear on the compound is
maximised.

The curatives are added late in the cycle to minimise their residence time since the mixer heats
up due to friction. In some cases the curatives will be added in a second stage, either in the mixer
or mill, to avoid starting the cure process or reducing its efficiency.

Some polymers such as EPDM do not need the initial softening and can be mixed "upside
down", with some of the fillers and oils added first, before the cycle continues as normal.

The material is dumped from the mixer at a pre-set temperature and/or energy value to ensure consistency of the final compound. Again this full cycle is recorded in the mixing computer.

The compound is then milled for initial cooling and to ensure homogeneity with soft compounds. A secondary cooling takes place in cooling racks or a specialised take-off unit which can apply anti-tack as needed.

6.4.4　Quality checking

The initial stage of quality checking to ensure the material meets the required standards normally includes the following tests.

（1）Cure characteristics

How an elastomer cures over time is measured on a rheometer. As the compound cures between the hot platens it becomes stiffer. This is measured via a strain gauge connected to an oscillating rotor in contact with the elastomer. The resistance of the material to the oscillating motion is plotted on a graph against time, known as a rheograph（Figure 6.12）, which enables the moulding characteristics to be predicted.

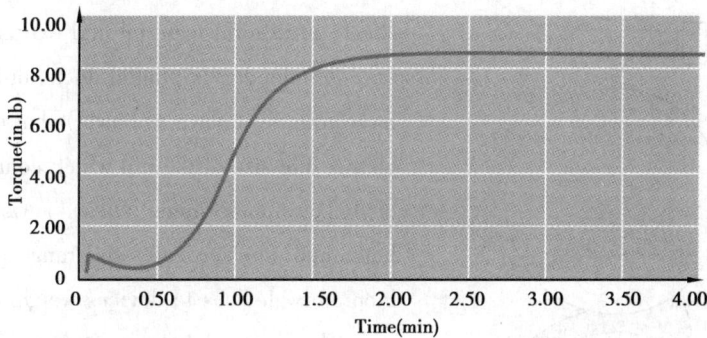

Figure 6.12　Typical rheometer trace showing progress of cure against time

（2）Hardness

Testing the hardness of the compound using an indentor provides a check that the correct levels of filler have been incorporated. The hardness should normally fall within +5 and −4 IRHD of the specified value. The strict process and quality control regimes in place at James Walker manufacturing sites however, ensures any variation in compound hardness is minimal and far below the industry standard range outlined here.

（3）Density

This is a measure of the weight per unit volume. It gives an indication of whether the correct quantities of ingredients have been added.

（4）Weight

This test checks the weight of the compound after leaving the mixer and compares it to the input weight of the mix to confirm all the constituents have been added.

6.4.5　Manufacturing

The manufacturing of products in elastomers involves a number of often complex operations to turn the raw material sheets into a finished product suitable for use. The operation can be spilt into three distinct areas of activity:

①Material preparation: This includes all the operations up to the point of moulding.

②Moulding: This includes turning the material into a cross-linked product.

③Post moulding operations: This involves finishing the product and ensuring that it meets all the necessary quality requirements.

(1)Material preparation

The elastomer from the mixer is normally available either as a sheet of predetermined thickness or split into rolls of material of known thickness and width. The latter option is often used where the material is fed directly into an injection moulding machine. A number of options are available in the development of the pre-form (blank) to be used in the moulding process. These include calendered sheet, strip form material and extrusion.

A calender is similar to a mill and has two or more rollers (known as bowls) that can be adjusted to change the size of the nip which controls the thickness of the elastomer sheet. These bowls can be mounted horizontally or vertically and range in size from small laboratory devices to devices weighing several tonnes.

The material from the mixer is fed between the nips on the calender and pulled away from the bowl by a manual or mechanical device. The desired sheet thickness is achieved by adjusting the nips.

The calender process allows a high degree of control on the thickness of the rubber sheet (Figure 6.13). This sheet is generally then used either to stamp a shape for placing into a mould in the next process, or to manufacture cross-linked elastomer sheeting from which gaskets or other finished products can be cut.

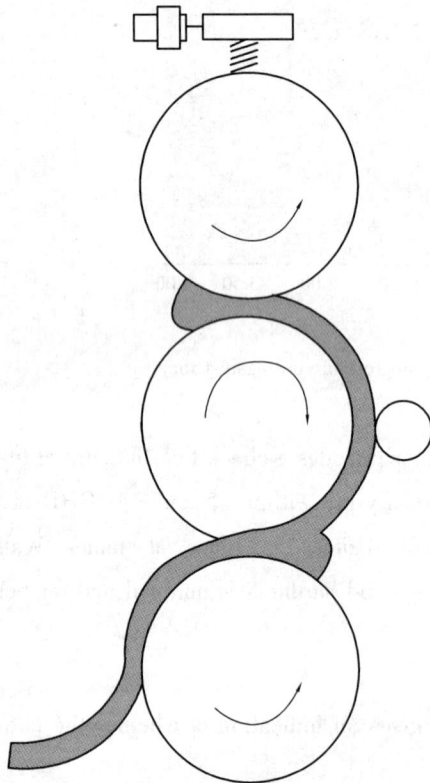

Figure 6.13　Typical calender

If the elastomer is to be fed directly into an injection moulding machine, the sheet from the mixing stage can be slit to create strips of elastomer. These are then fed directly into the screw feed of the injection moulding machine.

There are two main types of extruders: screw and ram.

①Screw extruders: Screw extruders have a screw housed within a barrel, with the screw turned by mechanical means (Figure 6.14). The elastomer is first fed into the barrel via a hopper and then forced down the barrel by the screw whilst heat is added (created by the shearing action and via the heated barrel and screw). At the end of the barrel, in the extruder head, is a die through which the material is forced out (Figure 6.15). The die is profiled to produce elastomer shaped for the next stage of processing.

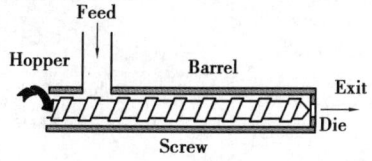

Figure 6.14   Screw extruder

②Ram extruders: For a ram extruder the elastomer needs to be rolled and warmed, usually by placing it in a bath of hot water or taking it directly from the mill/calender. This roll is then placed into the cylinder housing the ram. The head of the extruder containing the die is then locked in place at the front of the extruder and the ram traversed forward, forcing the material out of the die orifice(Figure 6.16). When the material exits the die it can either be pulled off in lengths or cut to length/weight by a rotating blade affixed to the front of the machine. For most materials (silicone being an exception) the extruder cylinder and head are heated.

Figure 6.15   Screw extruder for the manufacture of "O" rings

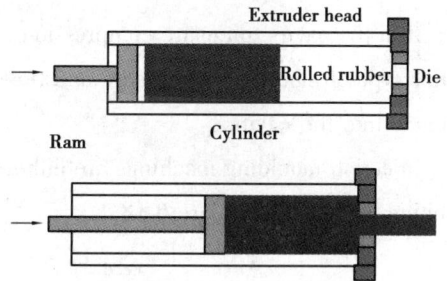

Figure 6.16   Ram extruders

In some cases, the extrusion profile created is the finished product and needs to be cross-linked to retain its shape. This, for example, is the process used for the manufacture of windscreen wiper blades. For products manufactured using this technique, the extrusion is cured either as it exits the machine through a hot box, or by other means following the extrusion process, such as autoclaving.

(2)Moulding

Three principal moulding techniques are used to manufacture elastomer products: compression moulding, injection moulding and transfer moulding.

Compression moulding describes the forming process in which an elastomer profile is placed directly in a heated mould, then softened by the heat, and forced to conform to the shape of the mould as the press closes the mould(Figure 6.17).

The presses are mostly hydraulically driven and can be either upstroking, where the lower platen moves up and the upper platen is fixed, or downstroking, where the upper platen is driven

169

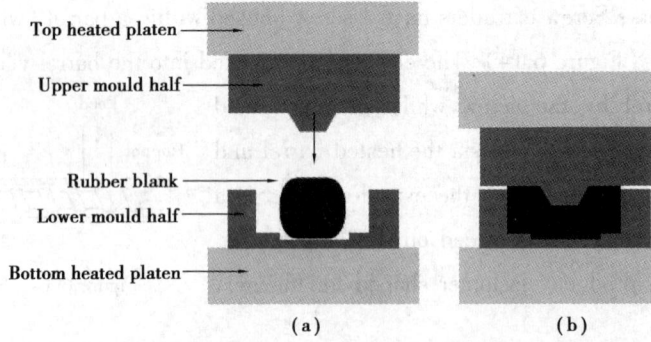

Figure 6.17 Open mould (a) with elastomer blank placed in cavity,
and (b) mould closed forming the finished product profile.

downwards and the lower platen remains fixed.

The advantages of compression moulding are the lower cost of the moulds, the large sizes of mouldings possible and the relatively quick changeover between different moulds.

The main disadvantage is output, as they are generally loaded and unloaded manually and the elastomer is often placed into the cavity "cold" so cure times are longer. Some difficulties that can occur are positioning the blank in the cavity and the "flash" that results from the additional material placed in the cavity to ensure compression in the cavity when the mould shuts. Another disadvantage of this type of moulding is the care and time required to manufacture the blank (weight and profile) to place into the cavity.

Injection moulding machines are either vertically or horizontally run. An example of a horizontal machine is shown in Figure 6.18.

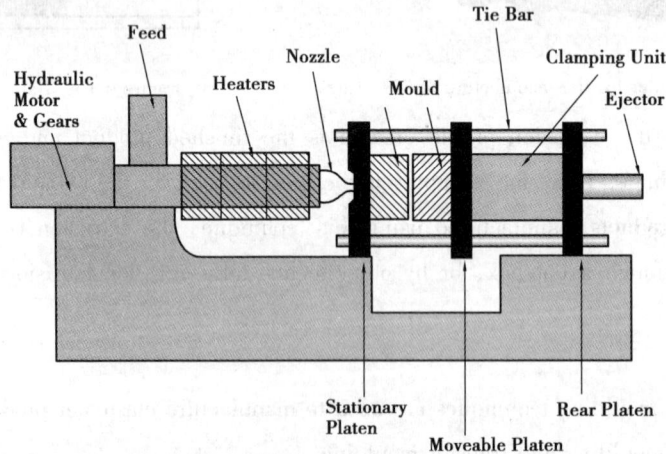

Figure 6.18 Horizontal injection moulding machine

Injection moulding is a process where heated elastomer is injected into a closed cavity via a runner system. Uncured elastomer is fed into the injection cylinder where it is preheated and accurately metered into the mould. This is done by controlling the pressure, injection time and temperature(Figure 6.19).

170

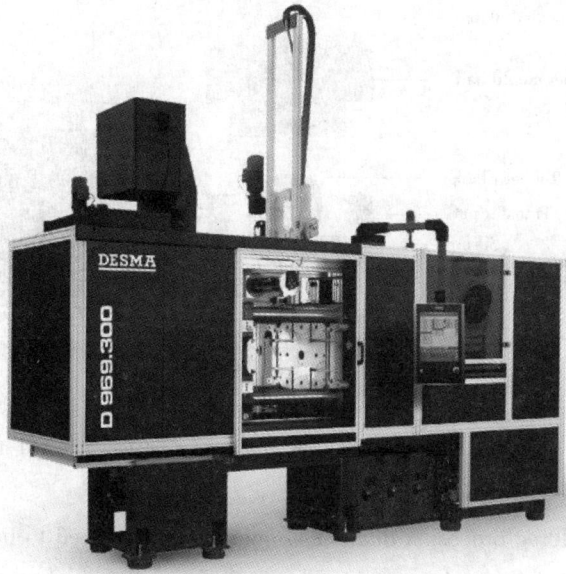

Figure 6.19   Rubber injection moulding machine

The advantages of injection moulding are its suitability for moulding delicate parts, shorter cycle times compared with compression moulding, the high levels of automation that can be introduced in the process and lower levels of flash since the mould is shut when the material is injected.

The main disadvantages of injection moulding are the costs of the tool, the longer changeover times resulting from the more complex tooling and the waste of material in the runner system where a hot runner system is employed (for thermoset materials). Material waste also occurs when jobs are run sequentially with either differing materials or different colours, which requires extensive purging of the machine. In the case of liquid silicone rubber (LSR), the injection machine can also be used to mix the two LSR constituents before injection into the mould.

Transfer moulding is a combination of injection moulding and compression moulding and takes place on a compression press(Figure 6.20). Elastomer of set weight is placed in the transfer pot, and, as with compression moulding, the pot is closed by the press forcing the elastomer down the sprues and into the cavity. A small amount of excess material flows out of the cavity through vents, with other excess material lying in the sprue grooves and a mat of material left in the transfer pot.

The advantage of this process over conventional compression moulding is the ability to form delicate parts and to mould parts having inserts requiring specific positioning within the product. It also has simpler blank requirements and faster cure times, since the elastomer heats up quickly as it is transferred from the pot to the cavity.

The main disadvantages are the additional cost of the tooling, the additional waste material due to the pot and sprue, and difficulties that can be experienced when transferring high hardness or high molecular weight materials.

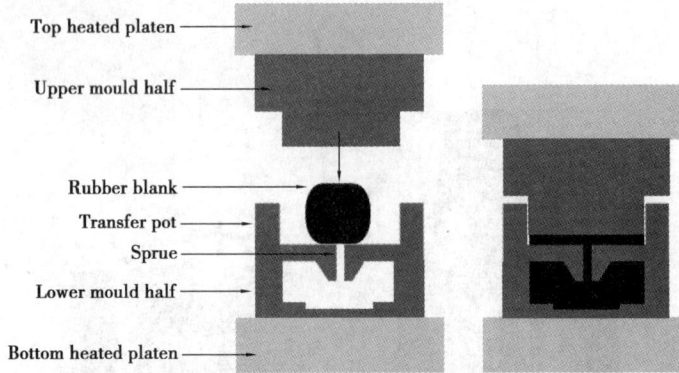

Figure 6.20   Transfer moulding process showing mould open (left) with the elastomer blank
in the transfer pot, and mould shut (right) with the elastomer injected into the cavity.

(3) Post moulding operations

Following the moulding process a number of operations may need to be performed to finish the
product. These include post-curing, removal of flash and injection sprues, other trimming
requirements, testing, etching, inspecting and packaging the product.

Various trimming techniques are available depending on the size and shape of the component
and the type of elastomer used (Figure 6.21). They include cryogenic, where the part is cooled
below its glass transition temperature and tumbled and blasted with beads to remove the flash,
buffing using abrasive wheels and belts, cutting the finished shape using cutting dies, formers,
knives and in some cases scissors, using lathes to trim and chamfer components, and rota finishing
where components are rotated amongst abrasive stones or other abrasive media.

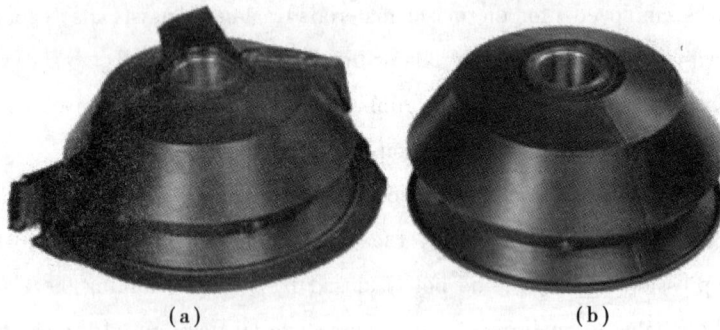

(a)   (b)

Figure 6.21   Component (a) before and (b) after trimming

For many heat resistant elastomers, such as fluorocarbon and silicone materials, it is necessary
to supplement the press cure with an oven post-cure to eliminate residues from peroxides and
complete the curing process.

With some materials a post-cure is required in an autoclave. This is a device which generally
uses steam to post-cure the components under pressure. An additional benefit of this type of post-
curing is that it provides a comprehensive "wash" of the products and is often used to finish
products for food or pharmaceutical applications.

172

Another use of autoclaves is to cure products (not to be confused with post-curing, as in this instance it is the primary curing process) that are too big or unsuitable to be moulded. Examples of such products are extrusions and sheetings.

In some cases the products manufactured need to be non-destructively tested to ensure they meet the required specifications. This normally takes the form of a hardness test.

Destructive testing of representative samples is also often carried out (compression set and immersion testing).

For some applications, the elastomer product requires to be etched to provide identification of its origin for branding purposes or other customer requirement.

One of the final steps in the manufacturing process is finished part inspection. This can be carried out by hand, with an inspector visually examining and measuring the products. Alternatively, for reasonably simple components, inspection can be performed by machine, using a contact or non-contact system (Figure 6. 22), in some cases working fully autonomously. Autonomous measurement is particularly suited to high volume production runs for products such as "O" rings.

Figure 6.22 Non-contact inspection

The finished products need to be packaged appropriately before shipping, sprotection from dirt and dust, etc, some components may need to be sealed against moisture or contamination by other fluids, or protected against UV light.

6.5　Products and applications of rubbers [6-11]

6.5.1　Natural rubber

Natural rubber (NR) is an elastomer with a basic monomer of cis-1,4-isoprene. It is made by processing the sap of the rubber tree (Figure 6.23) (i.e., *Hevea brasiliensis*) with steam, and compounding it with vulcanizing agents, antioxidants, and fillers. The white sticky sap of the milkweed is also a latex. Latex will turn into a rubbery mass within 12 hours after it is exposed to the air. The latex protects the tree or plant by covering the wound with a rubbery material like a bandage.

Figure 6.23　A rubber plantation in Viet Nam

Natural rubber is widely used for applications requiring abrasion or wear resistance, electric resistance and damping or shock absorbing properties such as large truck tyres, off-the-road giant tyres and aircraft tyres. It is chemically resistant to acids, alkalis and alcohol. However, it does not do well with oxidizing chemicals, atmospheric oxygen, ozone, oils, petroleum, benzene, and ketones.

NR macromolecules are susceptible to fracture on shearing. High shearing stresses and oxygen promote the rate of molecular chain scission.

Several modified natural rubbers are available commercially. Some examples are:

①Deproteinized, to reduce water adsorption, e.g., in electrical applications where maximum resistivity is required.

②Skim rubber, a high-protein, fast curing product used in cellular foams and pressure sensitive adhesives.

③Superior processing, in which ordinary and vulcanized latices are blended in about an 80 : 20 ratio before coagulation. Unfilled or lightly filled compounds made with superior processing NR give smoother and less swollen extrudates compared to those prepared from regular NR.

④ Isomerized, prepared by milling NR with butadiene sulfone, resulting in cis/trans isomerization which inhibits crystallization.

⑤Epoxidized, an oil resistant rubber, which retains the ability to strain crystallize.

Synthetic polyisoprene (IR) is produced both anionically and by Ziegler-Natta polymerization. The former material has up to 95% cis-1,4 microstructure, while the latter may be as much as 98% stereoregular. Even though the difference in stereoregularity is small, Ziegler-Natta IR is substantially more crystallizable. However, both types of IR have less green strength and tack than NR. IR compounds have lower modulus and higher breaking elongation than similarly formulated NR compositions. This is due, at least in part, to less strain-induced crystallization with IR, especially at high rates of deformation.

6.5.2   Butadiene rubber

Like isoprene, Butadiene rubber (BR) can be synthesized anionically or via Ziegler-Natta catalysis. Cold emulsion BR is also available. Anionic BR, prepared in hydrocarbon solvent, contains about 90% 1,4 structure and 10% 1,2 (i.e., vinyl). The vinyl content can be increased by adding an amine or ether as co-solvent during polymerization. The 1,4 structure is an approximately equal mix of cis and trans. Because it consists of mixed isomers, anionically prepared BR does not crystallize. Emulsion BR has a mostly trans microstructure and also does not crystallize. On the other hand, the Ziegler-Natta product has a very high cis content and can crystallize. The T_g of low-vinyl BRs is about -100 ℃, among the lowest of all rubbers, while that of high-vinyl BRs can reach 0 ℃. Low-vinyl BRs are highly resilient and are often blended with SBR, NR, and IR to make tire treads with good abrasion resistance. Unlike NR, BR is resistant to chain scission during mastication.

6.5.3   Styrene butadiene rubber

Styrene butadiene rubber (SBR) denotes a copolymer of styrene and butadiene, typically containing about 23% styrene, with a T_g of approximately -55 ℃. It is the most widely used synthetic elastomer, with the largest volume production. It is synthesized via free-radical polymerization as an emulsion in water, or anionically in solution. In emulsion polymerization, the emulsifier is usually a fatty acid or a rosin acid. The former gives a faster curing rubber with less tack and less staining. The molecular weight is controlled (to prevent gelation) by mercaptan chain transfer agents. When polymerization is complete, coagulation of the emulsion is carried out with salt, dilute sulfuric acid, or an alum solution. Alum coagulation results in a rubber with the highest electrical resistivity.

When emulsion polymerization is carried out at an elevated temperature (~ 50 ℃), the rate of radical generation and chain transfer is high, and the polymer formed is highly branched. To overcome this, polymerization is carried out at low temperature (~ 5 ℃), producing "cold" emulsion SBR, with less branching, and giving stronger vulcanizates. A common initiator for anionic polymerization is butyl lithium. The vinyl butadiene content, and hence Tg, of SBRs polymerized in solution are increased by increasing solvent polarity. In comparison with emulsion polymers, the molecular weight distribution of anionically prepared SBR is narrow, and because the chain ends are "living," i.e., they remain reactive after polymerization, the molecules can be functionalized or coupled. For example, SBR macromolecules can be amine-terminated to provide increased interaction with carbon black, or coupled with $SnCl_4$ to give star-shaped macromolecules that break upon mastication in the presence of stearic acid to yield a material with lower viscosity. Solution SBR is also purer than emulsion SBR, because of the absence of emulsion residues. But, when compared at similar number-average molecular weights, emulsion SBRs are more extensible in the uncured (so-called "green")state than anionic SBRs (Figure 6.24).

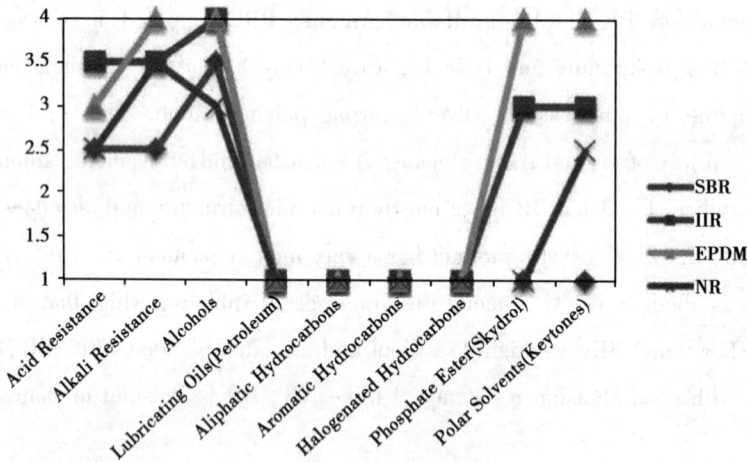

Figure 6.24   Schematic representation of hydrocarbon elastomer properties
(1 Poor; 2 Fair; 3 Good; 4 Excellent)

6.5.4   Chloroprene rubber

Polychloroprene is an emulsion polymer of 2-chlorobutadiene and has a T_g of about −50 ℃. The electron-withdrawing chlorine atom deactivates the double bond towards attack by oxygen and ozone and imparts polarity to the rubber, making it resistant to swelling by hydrocarbons. Compared to general-purpose elastomers, Chloroprene rubber (CR) has superior weatherability, heat resistance, flame resistance, and adhesion to polar substrates, such as metals. In addition, CR has lower permeability to air and water vapor.

The microstructure of CR is mostly trans-1,4 and homopolymer grades crystallize upon standing or straining, even though they are not as stereoregular as NR. Apparently, C-Cl dipoles enhance interchain interaction and promote crystallization. Copolymer grades of CR crystallize less or not at all. Applications include wire, cable, hose, and some mechanical goods.

6.5.5   Butyl rubber

Butyl rubber(BR) is a copolymer of isobutylene with a small percentage of isoprene to provide sites for curing. IIR has unusually low resilience for an elastomer with such a low T_g (about -70 ℃). Because IIR is largely saturated, it has excellent aging stability. Another outstanding feature of butyl rubber is its low permeability to gases. Thus, it is widely used in inner tubes and tire innerliners.

The development of halogenated butyl rubber (halobutyl) in the 1950s and 1960s greatly extended the usefulness of butyl by providing much higher curing rates and enabling co-vulcanization with general purpose rubbers such as natural rubber and styrene-butadiene rubber (SBR). These properties permitted development of more durable tubeless tires with the air retaining innerliner chemically bonded to the body of the tire (Figure 6.25). Tire innerliners are by far the largest application for halobutyl today. Both chlorinated (chlorobutyl) and brominated (bromobutyl) versions of halobutyl are commercially available. In addition to tire applications, butyl and halobutyl rubbers' good impermeability, weathering resistance, ozone resistance, vibration dampening, and stability make them good materials for pharmaceutical stoppers, construction sealants, hoses, and mechanical goods. The total annual demand for butyl polymers is

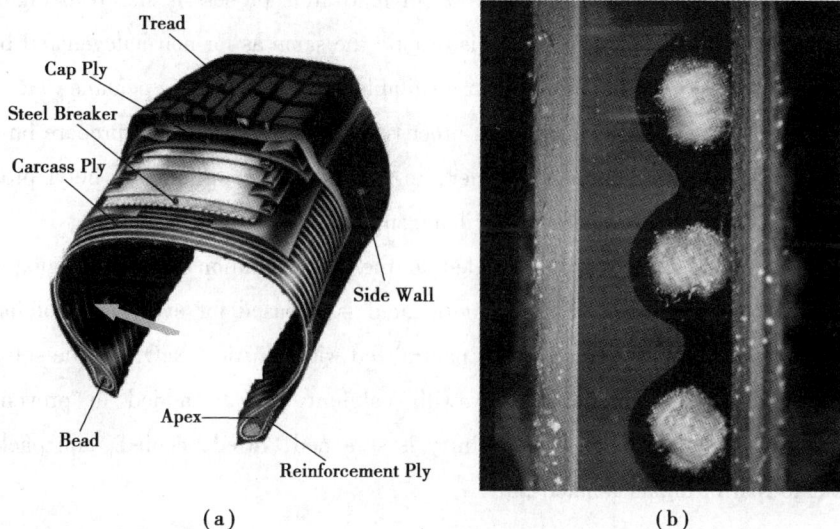

(a)

(b)

Figure 6.25   Schematic diagram of tire and Photomicrograph of tire innerliner

~650 000 metric tons.

Butyl Rubber typically contains about 98% polyisobutylene with 2% isoprene distributed randomly in the polymer chain. To achieve high molecular weight, the reaction must be controlled at low temperatures (−90 to −100 ℃). The reaction is highly exothermic. The most commonly used polymerization process uses methyl chloride as the reaction diluent and boiling liquid ethylene to remove the heat of reaction and maintain the needed temperature. It is also possible to polymerize butyl in alkane solutions and in bulk reaction. A variety of Lewis acids can be used to initiate the polymerization. The molecular weight of butyl is set primarily by controlling the initiation and chain transfer rates.

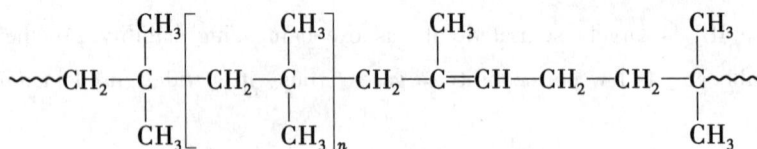

$$\sim\!\!\sim\!\!CH_2-\underset{\underset{CH_3}{|}}{\overset{\overset{CH_3}{|}}{C}}\left[CH_2-\underset{\underset{CH_3}{|}}{\overset{\overset{CH_3}{|}}{C}}\right]_n CH_2-\underset{CH_3}{\overset{\overset{CH_3}{|}}{C}}\!=\!CH-CH_2-CH_2-\underset{\underset{CH_3}{|}}{\overset{\overset{CH_3}{|}}{C}}\!\sim\!\!\sim$$

In the most widely used manufacturing process, a slurry of fine particles of butyl rubber (dispersed in methyl chloride) is formed in the reactor. The methyl chloride and unreacted monomers are flashed and stripped overhead by addition of steam and hot water, and then they are dried and purified in preparation for recycle to the reactor. Slurry aid (zinc or calcium stearate) and antioxidant are introduced to the hot water/polymer slurry to stabilize the polymer and prevent agglomeration. Then the polymer is screened from the hot water slurry and dried in a series of extrusion dewatering and drying steps. Fluid bed conveyors and/or airvey systems are used to cool the product to acceptable packaging temperature. The resultant dried product is in the form of small "crumbs", which are subsequently weighed and compressed into 75 lb. bales for wrapping in PE film and packaging. The commercial butyl slurry polymerization process is shown in Figure 6.26.

The polymerization process for halobutyl is exactly the same as for non-halogenated butyl. Prior to halogenation, the butyl must be dissolved in a suitable solvent (hexane, pentane, etc...) and all unreacted monomer removed. Several different processes are currently used to prepare butyl solution for halogenation. Either reactor effluent polymer, in-process rubber crumb, or butyl product bales may be dissolved in solvent in preparation for halogenation.

Bromine liquid or chlorine vapor is added to the butyl solution in highly agitated reaction vessels. One mole of hydrobromic or hydrochloric acid is released for every mole of halogen that reacts, therefore the reaction solution must be neutralized with caustic (NaOH). The solvent is then flashed and stripped by steam/hot water, with calcium stearate added to prevent polymer agglomeration. The resultant polymer/water slurry is screened, dried, cooled, and packaged in a process similar to that of unhalogenated butyl.

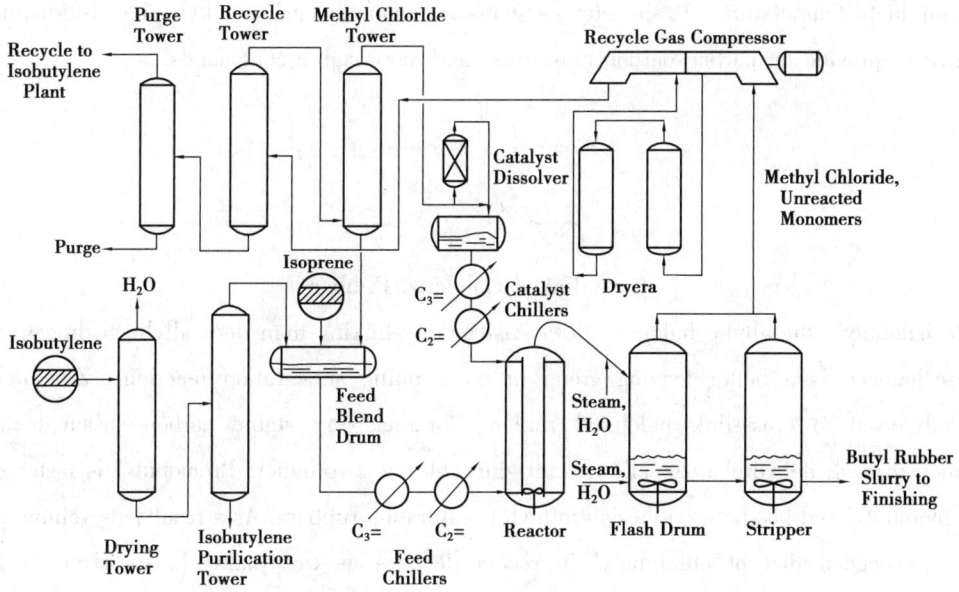

Figure 6.26   Commercial butyl slurry polymerization process

Most Abundant Halobutyl Isomer   Minor Halobutyl Isomers

Like other rubbers, for most applications, butyl rubber must be compounded and vulcanized (chemically cross-linked) to yield useful, durable end use products. Grades of Butyl have been developed to meet specific processing and property needs, and a range of molecular weights, unsaturation, and cure rates are commercially available. Both the end use attributes and the processing equipment are important in determining the right grade of Butyl to use in a specific application. The selection and ratios of the proper fillers, processing aids, stabilizers, and curatives also play critical roles in both how the compound will process and how the end product will behave.

Care must be taken when processing Halobutyl that premature dehydrohalogenation does not occur due to high temperature. Stabilizers (calcium stearate alone for chlorobutyl, supplemented with an epoxy compound such as epoxidized soybean oil in the case of bromobutyl) are required to prevent dehydrohalogenation during processing.

Elemental sulfur and organic accelerators are widely used to cross-link butyl rubber for many applications. The low level of unsaturation requires aggressive accelerators such as thiuram or thiocarbamates. The vulcanization proceeds at the isoprene site with the polysulfidic cross links attached at the allylic positions, displacing the allylic hydrogen. The number of sulfur atoms per cross-link is between one and four or more. Cure rate and cure state (modulus) both increase if the diolefin content is increased (higher unsaturation). Sulfur cross-links have limited stability at

179

sustained high temperature. Resin cure systems (commonly using alkyl phenol-formaldehyde derivatives) provide for carbon-carbon cross-links and more stable compounds.

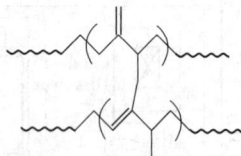

Cross-linked Halobutyl Rubber

In halobutyl, the allylic halogen allows easier cross-linking than does allylic hydrogen alone, because halogen is a better leaving group in nucleophilic substitution reactions. Zinc oxide is commonly used to cross-link halobutyl rubber, forming very stable carbon-carbon bonds by alkylation through dehydrohalogenation, with zinc chloride byproduct. Bromobutyl is faster curing than chlorobutyl and has better adhesion to high unsaturation rubbers. As a result, its volume growth rate has exceeded that of chlorobutyl in recent decades as tire plants have driven to higher productivity operation.

Butyl (and its primary derivative, halobutyl) is and will continue to be a high value polymer particularly well suited for its primary application of air retention in tires. Its unique combination of properties (excellent impermeability, good flex, good weatherability, co-vulcanization with high unsaturation rubbers, in the case of halobutyl) make it a preferred material for this application. As miles driven, tire size, and market sensitivity to pressure retention are all increasing, the demand for butyl rubber (specifically halobutyl) will continue to grow.

6.5.6 Special synthetic rubber

(1) Acrylonitrile-butadiene rubber

Acrylonitrile-butadiene rubber (NBR), also termed nitrile rubber, is an emulsion copolymer of acrylonitrile and butadiene. Acrylonitrile content varies from 18 to 50%. Unlike CR, polarity in NBR is introduced by copolymerization with the polar monomer, acrylonitrile, which imparts excellent fuel and oil resistance. Nith increased acrylonitrile content, there is an increase in T_g, reduction in resilience, lower die swell, decreased gas permeability, increased heat resistance, and increased strength. Because of unsaturation in the butadiene portion, NBR is still rather susceptible to attack by oxygen and ozone. Aging behavior can be improved by blending with small amounts of polyvinyl chloride. Nitrile rubber is widely used for seals and fuel and oil hoses.

$$H_2C \quad \overset{H_2}{\underset{}{C}} \quad R$$

$$\text{trans 1,4 BD} \qquad \text{Acrylo-Nitrile} \qquad \text{cis 1,4 BD} \qquad \text{trans 1,4 BD} \qquad \text{1,2 BD}$$

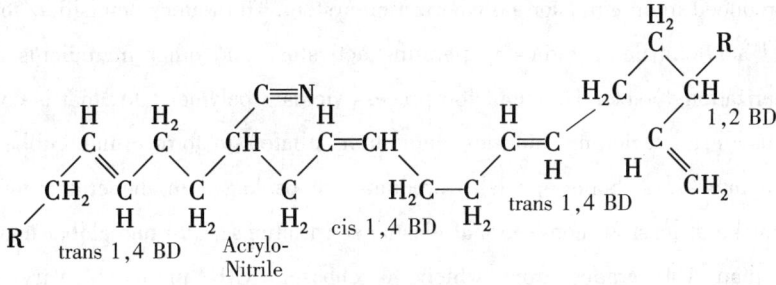

NBR Structure

NBR is commonly considered the workhorse of the industrial and automotive rubber products industries. NBR is actually a complex family of unsaturated copolymers of acrylonitrile and butadiene. By selecting an elastomer with the appropriate acrylonitrile content in balance with other properties, the rubber compounder can use NBR in a wide variety of application areas requiring oil, fuel, and chemical resistance. In the automotive area, NBR is used in fuel and oil handling hose, seals and grommets, and water handling applications. With a temperature range of −40℃ to +125 ℃, NBR materials can withstand all but the most severe automotive applications. On the industrial side NBR finds uses in roll covers, hydraulic hoses, conveyor belting, graphic arts, oil field packers, and seals for all kinds of plumbing and appliance applications. Worldwide consumption of NBR is expected to reach 368 000 metric tons annually by the year 2005.

Like most unsaturated thermoset elastomers, NBR requires formulating with added ingredients, and further processing to make useful articles. Additional ingredients typically include reinforcement fillers, plasticizers, protectants, and vulcanization packages. Processing includes mixing, pre-forming to required shape, application to substrates, extrusion, and vulcanization to make the finished rubber article. Mixing and processing are typically performed on open mills, internal mixers, extruders, and calenders. Finished products are found in the marketplace as injection or transfer molded products (seals and grommets), extruded hose or tubing, calendered sheet goods (floor mats and industrial belting), or various sponge articles. Figure 6.27 shows some typical molded and extruded NBR products.

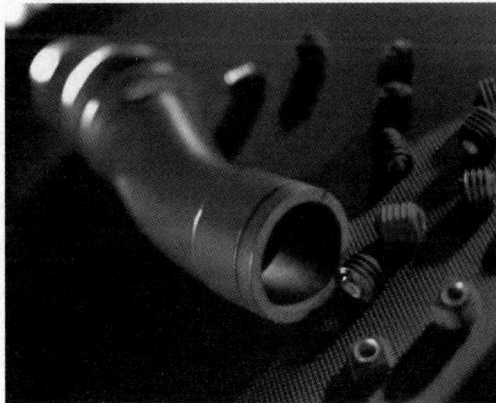

Figure 6.27   Typical finished NBR articles

NBR is produced in an emulsion polymerization system. The water, emulsifier/soap, monomers (butadiene and acrylonitrile), radical generating activator, and other ingredients are introduced into the polymerization vessels. The emulsion process yields a polymer latex that is coagulated using various materials (e.g., calcium chloride, aluminum sulfate) to form crumb rubber that is dried and compressed into bales. Some specialty products are packaged in the crumb form. Most NBR manufacturers make at least 20 conventional elastomer variations, with one global manufacturer now offering more than 100 grades from which to choose. NBR producers vary polymerization temperatures to make "hot" and "cold" polymers. Acrylonitrile (ACN) and butadiene (BD) ratios are varied for specific oil and fuel resistance and low temperature requirements. Specialty NBR polymers which contain a third monomer (e.g., divinyl benzene, methacrylic acid) are also offered. Some NBR elastomers are hydrogenated to reduce the chemical reactivity of the polymer backbone, significantly improving heat resistance (see HNBR product summary). Each modification contributes uniquely different properties. Figure 6.28 shows the typical NBR manufacturing process.

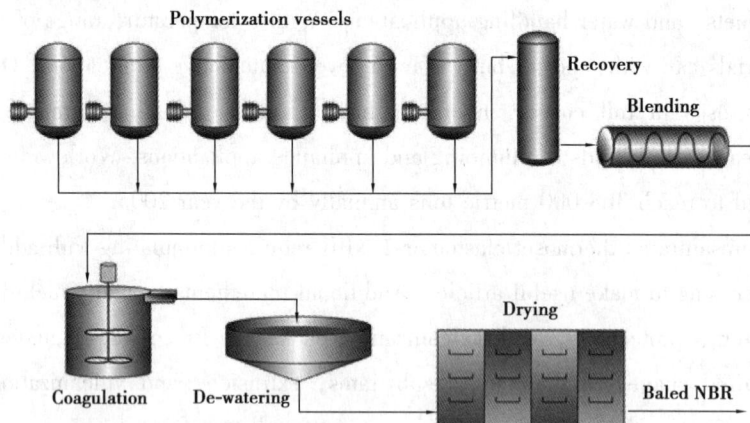

Figure 6.28   NBR manufacturing process

The ACN content is one of two primary criteria defining each specific NBR grade. The ACN level, by reason of polarity, determines several basic properties, such as oil and solvent resistance, low-temperature flexibility/glass transition temperature, and abrasion resistance. Higher ACN content provides improved solvent, oil and abrasion resistance, along with higher glass transition temperature.

Mooney viscosity is the other commonly cited criterion for defining NBR. The Mooney test is reported in arbitrary units and is the current standard measurement of the polymer's collective architectural and chemical composition. The Mooney viscosity provides data measured under narrowly defined conditions, with a specific instrument that is fixed at one shear rate. Mooney viscosity of polymers will normally relate to how they will be processed. Lower Mooney viscosity materials (30~50) will be used in injection molding, while higher Mooney products (60~80) can be more highly extended and used in extrusion and compression molding.

(2) Silicone rubber

Silicone rubber's special features such as "Organosiloxanes Polymer" has been originated from its unique molecular structure that they carry both inorganic and organic properties unlike other organic rubbers. In other words, due to the Si-O bond of silicone rubber and its inorganic properties, silicone rubber is superior to ordinary organic rubbers in terms of heat resistance, chemical stability, electrical insulating, abrasion resistance, weatherability and ozone resistance etc…

With these unique characteristics, silicone rubber has been widely used to replace petrochemical products in various industries like aerospace, munitions industry, automobile, construction, electric and electronics, medical and food processing industry. Recently, these scopes of silicone application have been expanding at a great speed by the demand of industries that want more reliable elastomer.

Main characteristics of silicone rubber:

①Excellent high and low temperature resistance.

②Excellent electrical properties.

③Physiological inertness.

④Excellent weatherability.

⑤Oil resistance.

⑥Flame retardant.

Silicone rubber has siloxane bond (Si-O) of molecular structure as the main chains. While carbon bond, C-C, carries 84.9 kcal/mol, siloxane bond carries 106.0 kcal/mol. It shows that siloxane bond has greater capacity and stability. As a result, silicone rubber has better heat resistance, electric conductivity and chemical stability than any other ordinary organic rubbers. Siloxane bond's energetic stability is secured due to sharp difference between Si and O in terms of electro-negativity making Si-O to be closest to ionic bond.

Heat resistance of silicone rubber is the one of its most excellent properties and provides the basis for its creation. Silicone rubber is far better than organic rubbers in terms of heat resistance. At 150 ℃, almost no alterations of properties take places that it may be used semi permanently. Furthermore, silicone rubber withstands use for over 10 000 consecutive hours even at 200 ℃ and, if used for a shorter term, it may also be used at 300 ℃ as well. Boasting this excellent heat resistance, silicone rubbers are widely used to manufacture rubber components and parts used in high-temperature places.

Cold resistance of silicone rubber is the finest among organic rubbers. It provides a critical reason behind the creation of silicone rubbers. Natural and ordinary rubbers demonstrate significant changes in formation depending on temperatures. They become soft at high temperatures and hard at low temperatures so that they may not be able to used any more. While other organic rubbers may only be used up to −20 ℃ or −30 ℃, silicone rubber maintains its elasticity between −55 ℃ and −70 ℃. Some of the products even withstand temperatures as extremely low as under −100 ℃.

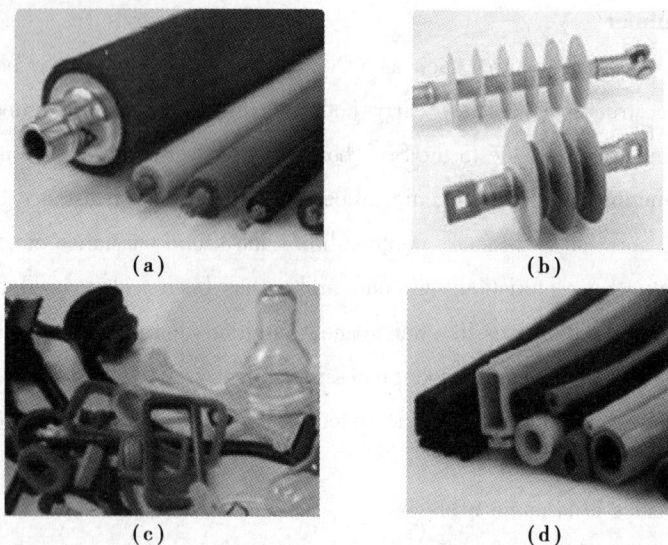

(a)　　　　　　　　　　　(b)

(c)　　　　　　　　　　　(d)

Figure 6.29　Products of silicone rubber

　　Silicone rubber has many applications as shovonin Figure 6.29 Silicone rubber is being used for insulation materials at high temperature with its superior insulation properties. It is particularly known for wide range in temperature and volume resistance between $10^{14} \Omega \cdot$ cm and $10^{16} \Omega \cdot$ cm. Silicone rubber experiences lowest change in performance in wet condition and is the best fit for being used as insulation materials. By adding special conductive fillers, conductive silicone may also be manufactured. In particular, silicone rubber is strongly resistant against corona discharge compares to others, while being widely used for insulation purposes in high voltage environments.

　　Silicone rubber has superb ozone resistance. Due to corona-discharged ozone, other organic rubbers become soften at a higher speed, but silicone rubber is rarely affected. Furthermore, even long-term exposures to UV rays, winds, or rain silicone rubber's physical properties will not be changed substantially.

　　Silicone rubber does not easily burn when in contact with a flame, but would burn out consistently once ignited. However, by adding a small amount of flame retardant, it may become flame retardant and self-extinguisher. Flame retardant silicone rubbers presently in use would scarcely produce toxic gas during combustion since they do not contain organic halogen compounds discovered in organic polymers.

　　Silicone rubber is inferior to ordinary organic rubber in oil resistance at room temperature. However, for automobiles or aircrafts that require high temperature resistance, it demonstrates higher performance. Even when in contact with automobile oil, silicone rubber does not inflate significantly by reason of swelling. It swells in nonpolar organic compounds such as benzene, toluene, and gasoline. But its materials do not disintegrate or dissolve unlike ordinary organic rubbers. If solvent is removed, it would be restored to the original conditions.

　　Silicone rubber is physiologically inert, and is thus used for baby nipple and stoppers in medical application. Silicone rubber is also very ideal elastomer for making swimming caps and goggles.

6.5.7   Thermoplastic elastomer [3]

Thermoplastic elastomers (TPEs) are unique synthetic compounds that combine some of the properties of rubber with the processing advantages of thermoplastics. Generally, they can be categorized into two groups: multi-block copolymers and blends. The first group is copolymers consist of soft elastomers and hard thermoplastic blocks, such as styrenic block copolymers (SBCs), polyamide elastomer block copolymers (COPAs), polyether ester-elastomer block copolymers (COPEs) and polyurethane-elastomer block copolymers (TPUs). TPE blends can be divided into polyolefin blends (TPOs) and dynamically vulcanized blends (TPVs). The world TPEs demand by types of TPE and by regions in year 2004, 2009 and 2014 are shown in Figure 6.30.

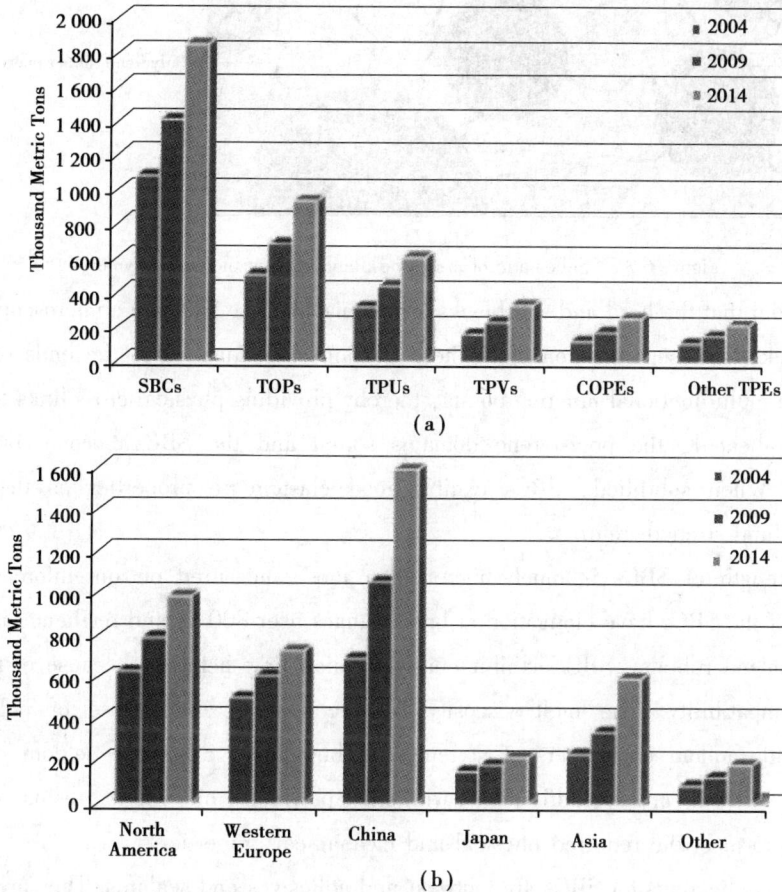

(a)

(b)

Figure 6.30   World thermoplastic elastomers demand in 2004,

2009 and 2014 by types of TPEs and by regions[12]

Styrenic block copolymers (SBCs) are the largest volume and lowest priced category of thermoplastic elastomers. SBCs are based on simple molecules (A-B-A type) that consist of at least three blocks, namely two hard polystyrene end blocks and one soft, elastomeric midblock. The midblock is typically a polydiene, either polybutadiene or polyisoprene, resulting in the well-known family of styrene-butadiene-styrene (SBS) and styrene-isoprene-styrene (SIS). Other SBCs which have been commercially successful include ethylene-butylene (SEBS), ethylene-propylene (SEPS), polyisobutylene (SIBS) and ethylene-ethylene-propylene (SEEPS). The structure of SBCs is shown schematically in Figure 6.31.

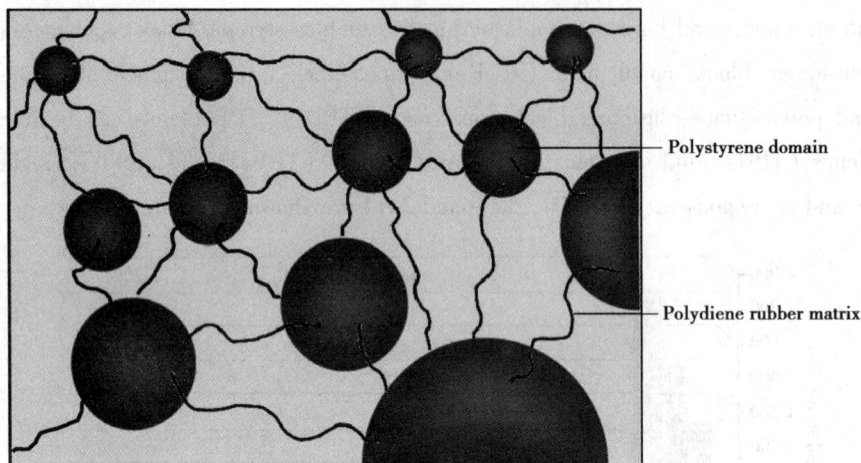

Figure 6.31　Schematic of a styrene-diene-styrene block copolymer

It is essential that the hard and soft blocks are immiscible, so that, on a microscopic scale, the polystyrene blocks form separate domains. These domains are attached to the ends of elastomeric chains and form multifunctional junction points, thereby providing physical cross links to the rubber. When they are heated, the polystyrene domains soften and the SBCs become processable as thermoplastics. When solidified, SBCs exhibit good elastomeric properties as the polystyrene domains reform and strength returns.

Tensile strength of SBCs is much higher than those measured on unreinforced vulcanized rubbers. Most of the SBCs have elongation at break ranges over 800 % and resilience is comparable to that of vulcanized rubbers. SBCs exhibit non-Newtonian flow behavior because of their extreme segmental incompatibility. The melt viscosity is much higher than those of polybutadienes, polyisoprene and random copolymers of styrene and butadiene. SBCs are seldom used as pure materials. Most of them can be readily mixed with other polymers, oil, fillers, resins, colorants and processing aids to meet the required physical and mechanical properties.

The major applications for SBCs are footwear and adhesives and sealants. They are also used in modifying the performance of asphalt for roofing and roads, particularly under extreme weather conditions. SBS block copolymers are among the most commonly modifiers for this application. Recently, it is proven that SEBS acts as a better modifier than SBS in improving the asphalts rutting resistance due to its double bond saturation which makes the SEBS more rigid than SBS. SBCs can also be compounded to produce materials that enhance grip, feel, and appearance in applications

186

such as toys, automotive, personal hygiene and packaging.

(1) **Thermoplastic elastomers based on polyamide**

Thermoplastic polyamide elastomers (COPAs) belong to the group of segmented block copolymer. They consist of multiblock copolymer structure with repeating hard and soft segments. The hard segments are polyamides which serve as virtual cross-links reducing chain slippage and the viscous flow of copolymer, whilst the soft segments are either polyethers or polyesters which contribute to the flexibility and extensibility of elastomers. Both segments are connected by amide linkages. The important members of COPAs are polyesteramides (PEAs), polyetheresteramides (PEEAs), polycarbonateesteramides (PCEAs) and polyether-block-amides (PE-b-As).

The properties of COPAs may vary according to such factors as the proportion of the hard and soft segments in the copolymer, chemical composition, molecular weight distribution, the method of preparation, and the thermal history of the sample that affects the degree of phase separation and domain formation. The hard segment controls the degree of crystallinity, crystalline melting point and mechanical strength while the soft segments determines the thermal oxidative stability, hydrolytic stability, chemical resistance and low temperature flexibility.

Most of the COPAs exhibit higher resistance to elevated temperature than any other commercial TPEs. They are also higher resistant to long-term dry heat aging without adding any heat stabilizers. Abrasion resistance of COPAs is comparable to that of TPUs. COPAs have excellent abrasion resistance, which is comparable to that of TPUs. The hardness is in the range from Shore 80A to Shore 70D by varying the content of hard and soft segments. The good insulation properties of COPAs make them suitable for low voltage applications and for jacketing. Other application areas for this material include conveyor and drive belt, footwear such as ski boots and sport shoes, automotive applications, electronics, hot melt adhesives, powder coatings for metals and impact modifiers for engineering thermoplastics.

Polyester amides constitute a peculiar of biodegradable family, due to the presence of both ester and amide groups that guaranties degradability. These biodegradable polymers are receiving great attention and are currently being developed for a great number of biomedical applications such as controlled drug delivery systems, hydrogels, tissue engineering, and other uses.

(2) **Thermoplastic polyether ester elastomers**

The polyether ester elastomers or copolyesters (COPEs) consist of sequence of hard and soft segments. The high melting blocks (hard segments) are formed by the crystalline polyester segments which are capable of crystallization and the rubbery soft segments are formed by the amorphous polyether segments with a relatively low glass transition temperature. At useful service temperature, the polyester blocks form crystalline domains embedded in rubbery polyether continuous phase. These crystalline domains act as physical cross-links. At elevated temperatures, the crystallites break down to yield a polymer melt, thus facilitating the thermoplastic processing.

COPEs are considered engineering thermoplastic elastomers because of their unusual combination of strength, elasticity and dynamic properties. They have a wide useful temperature range between the glass transition temperature (around −50 ℃) and melting point (around 200 ℃). These materials are elastic but their recoverable elasticity is limited to low strains. They have excellent dynamic performance and show resistance to creep. COPEs can be used in electrical applications for voltages 600 V and less. COPEs are resistant to oils, aliphatics and aromatic hydrocarbons, alcohols, ketones, esters and hydraulic fluids.

These materials, because of their high modulus and stiffness, have been used to replace some convention rubbers, PVC and other plastics in many applications. Uses of COPEs are reported in fuel tanks, quiet running gear wheels, hydraulic hoses, tubing, seals, gaskets, flexible couplings, wire and cable jacketing.

(3) Polyolefin-based thermoplastic elastomers

Polyolefin thermoplastic elastomers (TPOs) are an important part of the TPEs, which consist of polyolefin semi-crystalline thermoplastic and amorphous elastomeric components. TPOs are co-continuous phase system with the hard phase providing the strength and the soft phase providing the flexibility. The two most important processing methods of TPOs are injection moulding and extrusion. Others processing methods include calendaring, thermoforming, negative thermoforming and blow molding. TPOs ingredients generally include ethylene-propylene random copolymer (EPM), isotactic poly propylene (IPP), and the addition of other fillers and additives.

TPOs share with all TPEs the fundamental characteristics of having elastomeric properties, yet they process like a thermoplastic material. They are available in the hardness range from 60 Shore A to 70 Shore D. Their flexural modulus can range from 1 000 psi to 250 000 psi (6.9 MPa ~ 1 725 MPa). TPOs containing high amounts of elastomers are quite rubbery with high elongation at break values while TPOs containing high amounts of polyolefin undergo a yield at low elongation and there is little recovery after drawing. Most TPOs are resistant to ozone, unaffected by water or aqueous solutions of chemicals. TPOs are excellent electrical insulating materials with the dielectric strength of 500 V/mil and the volume resistivity at 23 ℃ and 50 % RH is 1.6×10^{16}.

Hard TPOs compound is often used for injection molded automotive interior or exterior fascia. Soft TPOs compound can be extruded into a sheet and thermoformed for large automotive part such as interior skin.

Collection of Exercises

一、简述题

1. Briefly explain the mechanism of elasticity of rubbers?

2. Compare the structural difference between natural rubber and synthetic polyisoprene.

3. Name the most important three commercial rubbers and explain why?

4. Give three examples of the application of rubber in industry.

5. Compare the structure and properties between SBR and SBS.

6. What are specialized rubbers? Explain the structure-properties relationship.

7. Give the advantages and disadvantages of thermoplastic rubbers.

8. How to choose a rubber according to given requirements.

9. Explain the function of all kind of additives in rubber, and name at least four common additives.

10. Please give the typical procedure of preparing sulphur vulcanized rubbers.

11. 聚乙烯和聚丙烯为通用塑料,但它们的共聚物却可以成为橡胶,为什么?

12. 举例说明热塑性弹性体的弹性原理。

13. 简述橡胶的加工工艺过程。

14. 简述橡胶作为材料的基本特点。

15. 硅橡胶具有优异的综合性能,它的硫化方式与通用橡胶有什么不同?

二、选择最合适的材料,并说明理由

1. 下列的哪一种材料最适合制造汽车的外胎?(　　　)

A. 天然橡胶　　　　　B. SBS　　　　　C. 丁基橡胶　　　　　　　　D. 丁腈橡胶

2. 下列的哪一种材料最耐臭氧?(　　　)

A. BR　　　　　　　B. NBR　　　　　C. ethylene-propylene rubber　D. natural rubber

3. 下列的哪一种材料最适合制作油密封垫?(　　　)

A. 天然橡胶　　　　　B. SBS　　　　　C. 丁基橡胶　　　　　　　　D. 丁腈橡胶

4. 下列的哪一种材料最适合耐溶剂性能最好?(　　　)

A. 天然橡胶　　　　　B. 丁腈橡胶　　　C. 硅橡胶　　　　　　　　　D. 氟橡胶

三、论述题

1. 列出目前主要商业化橡胶品种的玻璃化转变温度(Tg)的大小顺序,并从分子结构与性能的角度分析出现差别的原因及其相关应用情况。

2. 从结构、性能角度出发,分析氟橡胶和硅橡胶具有优异综合性能的主要原因,并根据其主要性能列举出相关应用领域。

REFERENCES

［1］www.iisrp.com/WebPolymers/00Rubber_Intro.pdf，History & Introduction of Rubber，IISRP 公司网站资料.

［2］http://www.jameswalkergroup.com/125yr_booklet.pdf，125 Years of the James Walker Group，James Walker 公司.

［3］Robert A. Shanks，Ing KongGeneral，Purpose Elastomers：Structure，Chemistry，Physics and Performance，Advanced Structured Materials 11，Springer-Verlag，Berlin Heidelberg，2013.

［4］http://www.jameswalker.biz/en/pdf_docs/148-elastomer-engineering-guide，Elastomer Engineering Guide，James Walker 公司.

［5］Rani Joseph，Practical Guide to Latex Technology，Smithers Rapra Technology，2013.

［6］Gary R. Hamed，Engineering with Rubber（chapter 2 Materials and Compounds），The University of Akron，Akron，Ohio 44325-3909，USA.

［7］James E. Mark，Burak Erman，Frederick R. Eirich. Science and Technology of Rubber，third edition，Elsevier Academic Press，2005.

［8］http://www.desma-usa.com/，DESMA 公司产品资料.

［9］http://lanxess.cn/cn/china-home/，郎盛公司产品技术资料.

［10］http://www.rubberdevelopment.com/，网页资料.

［11］Herman F. Mark et.al. Encyclopedia of Polymer Science and Engineering，3rd Edition，Wiley-Interscience，1987.

［12］Jiri，G.D.：Handbook of Thermoplastic Elastomer，pp. 161-177. William Andrew Publishing，New York（2007）.

CHAPTER 7
COATINGS AND ADHESIVES

7.1 Introduction

Paints and coatings occupy a prominent place in the cultural history of mankind. People have always been fascinated with colors and used paints to decorate and beautify themselves and their environment. Archaeologists found pigments and paint grinding equipment in Zambia (Southern Africa), thought to be between 350 000 and 400 000 years old. The oldest testimony of artistic activity was found in the south of France where some 30 000 years ago our prehistoric ancestors decorated the walls of their cave dwellings with stunning animal drawings. The ancient paints consisted of animal fat and colored earth or natural pigments such as ochre. Hence, they were based on the same principle as the paints that are used today—a binder and a coloring agent. Around 3 000 B.C., Egyptians began using varnishes and enamels made of beeswax, gelatin and clay and later protective coatings of pitch and balsam to waterproof their wooden boats. About 1 000 B.C., the Egyptians created varnishes from Arabic gum. Independently, the ancient Asian cultures developed lacquers and varnishes and by the 2nd century B.C., these were being used as coverings on a variety of buildings, artwork and furnishings in China, Japan and Korea. The ancient Chinese knew how to make black lacquer (the true predecessor of modern coatings) from the sap of the lacquer tree *Rhus vernicifera*. In India, the secrete of the lac insect *Coccus lacca* was used to produce a clear coat to beautify and protect wooden surfaces and objects. The early Greeks and Romans also relied on paints and varnishes, adding colors to these coatings and applying them on homes, ships, and artwork. It was not until the industrial revolution and ensuing mass production of linseed oil, however, that the production of modern house-paints began[1-4].

Over the decades, formulations of paints and coatings have become more and more sophisticated. Today, coatings not only protect and beautify the substrate. They also have functional properties: they are used as anti-skid surfaces, they can insulate or act as an electrical conductor, they can reflect or absorb light, etc. Paints and coatings play an indispensable role in our modern world and cover virtually everything we use from household appliances, buildings, cars, ships, airplanes to computers, microchips or printed circuit boards[5].

Coating materials are viscous liquid usually based on resins or oils, followed by adding or not adding pigments and extenders and then modulated by water or organic solvent. There are new kinds of coating materials which exist in solid form such as powder coating materials. No matter how the form varies(solid or liquid)from different kind of coating materials, they should be composed of at least two or three basic components. Those components are main film-forming substance, secondary film-forming substances and assistant film-forming substances. Coating material are liquid or solid polymer materials which can coat on the substrate and form tough and continuous films, with the old saying of oil paint. Coating material is usually used for decorating and protecting the coated surface. Some of coating materials have some special functions such as heat resistance, cold resistance and radiation proof, etc. Coating materials are widely used in aspects such as architecture, watercraft, vehicle and metal work.

Adhesive is a kind of substance that can integrate same or different kinds of materials compactly. Cementing refers to the technology that homogeneous or heterogeneous surfaces connected together by adhesives. It has characteristics like continuous stress distribution, lightweight, sealable, low temperature in most of the process. There are many ways to classify the adhesives. According to method of application, it can be classified into thermosetting adhesive, melt adhesive, adhesive cured at room temperature, pressure sensitive adhesive, etc. According to the application object, it can be classified into structure type, non-structure type or special gelatin. According to the morphology, it can be classified into water soluble type, emulsion type, solvent type and various kinds of solid type. Synthetic chemistry workers are more likely to classify the adhesives by chemical component of binder. With the development of science and technology and national economy, adhesives are already widely used in different industry and high-technology fields such as wood processing, construction engineering, textile printing and dyeing, shoes, leather, electronics, automobile, machinery, aerospace, biomedical and so on.

In order to understand the importance of the role adhesives play in the world, it was necessary to understand the history behind adhesives. The first observed adhesive could be dated back to 4 000 B.C., in which pre-historic tribes plugged broken pottery vessels with tree sap in which they stored foodstuffs in the coffins of dead people. Between 2 000 and 1 000 B.C., animal glue began to be used throughout civilization, as paintings, murals, and caskets contained glue in their construction. Artifacts from ancient Egypt, such as the tombs of pharaohs, were observed to be bonded or laminated with some form of animal glue.

The Greeks and Romans, approximately 2 000 years later, began to improve on this glue by incorporating various natural substances into adhesives to provide better bonding strength.

Ingredients such as egg whites, blood, bones, hide, milk, cheese vegetables, grains, beeswax, and tar were all used in various forms of manufacturing and artwork, such as ship construction and veneering and marquetry, in which thin sections of layers of wood were bonded together.

For the next several hundred years, adhesives became more widespread as furniture and cabinet makers incorporated adhesives into their work. Some of these makers can be recognized today, such as Chippendale and Duncan Phyfe. Adhesives also have played an important role in military history, as most weapons parts in the early part of the millennium were bonded solely with adhesives. Violins were laminated with a specialty adhesive, and violin makers today have yet to recreate the lamination process of the 1500 and 1600s.

In the 1700s, the adhesive industry really began to take off, as the first glue factory was constructed in Holland in which animal glues were manufactured from hides. In the late 1700s, patents began to be issued for glues and adhesives, as fish glue and adhesives using natural rubber, animal bones, fish, starch, and milk protein were all patented. By the start of the industrial revolution, the United States had several large glue-producing factories. As the 1900s progressed, the discovery of oil helped the adhesive industry take off in great proportion, as this led to the discovery of plastics. The introduction of Bakelite phenolic allowed adhesives using resin to be put on the market, and within the next 40 years, as new plastics and rubbers were being synthetically produced, the present day technology of adhesives were discovered. This development of plastics and elastomers has allowed the properties of adhesives to be changed and improved, such as flexibility, toughness, curing or setting time, temperature and chemical resistance[6].

7.2 Coatings

7.2.1 Alkyd coatings[7]

A modern coating is a complex mixture of components(Table 7.1). Oil paint, the oldest form of modern paint, uses a binder that is derived from a vegetable oil, such as obtained from linseed or soya bean. In alkyd paint, the binder is a synthetic resin, which is called an alkyd resin. The term "alkyd" was coined in the early days and originates from the AL in polyhydric Alcohols and the CID (modified to KYD) in polybasic acids. Hence, in a chemical sense the terms alkyd and polyester are synonymous. Commonly, the term "alkyd" is limited to polyesters modified with oils or fatty acids. A typical alkyd resin is prepared by heating for example linseed oil, phthalic acid anhydride and glycerol to obtain a fatty-acid containing polyester, as schematically shown Figure 7.1. Paints based on alkyd resin binders are usually solvent-borne paints, common solvents being white spirit(a mixture of saturated aliphatic and alicyclic C7-C12 hydrocarbons with a content of 15%-20% (by weight)of aromatic C7-C12 hydrocarbons), or xylene.

Table 7.1 **Typical coating composition**

Component	Weight-%
Binder	30
Organic solvent	27
Water	10
Pigments	19
Extenders	12
Additives	2

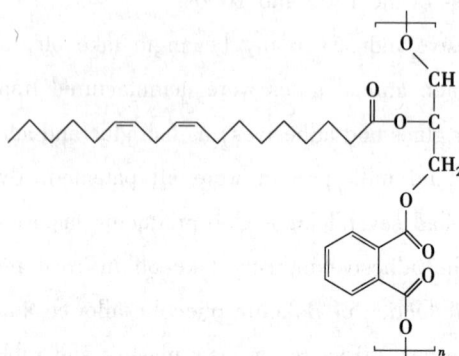

Figure 7.1   Example of an alkyd resin used as a binder compound in alkyd paint.
The fatty acid chain shown is linoleic acid.

Although European legislation drives paint development towards water-borne systems, in order to reduce the amount of VOC's(volatile organic compounds) in the atmosphere, solvent-borne paints often show a number of advantages over water-borne paints. Examples are: easier application properties, wider application and drying tolerance under adverse conditions(low temperature, high humidity) and a higher level of performance on difficult substrates, such as heavily stained or powdery substrates. As a result solvent-borne coatings will not be totally replaced by water-borne coatings in the foreseeable future, according to the paint industry. Further important components of alkyd paint are pigments and extenders. The pigment is the substance that gives the paint color. Pigments are derived from natural or synthetic materials that have been ground into fine powders. Extenders are inert pigments used to extend or increase the bulk of a paint. Extenders are also used to adjust the consistency of a paint and to let down colored pigments of great tinting strength. The last important category of components of alkyd paint comprises the additives. A large variety of coating additives is known, which have widely differing functions in a coating formulation. One of the most important groups of additives is that of the catalytically active additives, which includes the paint drying catalysts, or driers. Driers are metal soaps or coordination compounds which accelerate paint drying, thus shortening the total drying time. Without driers, the drying time of alkyd paint would be over 24 hours, which is clearly undesirable for most applications.

During the drying of alkyd paints several different stages can be identified. The first process is

194

the physical drying of the paint. In this process, the solvent evaporates and a closed film forms through coalescence of the binder particles. Then chemical drying(also called oxidative drying) occurs, a lipid autoxidation process, which means that the paint dries by oxidation of the binder compound with molecular oxygen from the air. Autoxidation will be discussed in detail in the next section. During the drying process four overlapping phases can be discriminated:

①Induction period

②Hydroperoxide formation

③Hydroperoxide decomposition into free radicals

④Polymerisation/cross-linking

The induction period is the time between application of the paint to a surface and the start of dioxygen uptake by the paint film. The induction period occurs because the effects of solvent, anti-skinning agent and natural anti-oxidants that may be present in the alkyd resin must be overcome before the drying process can begin. Autoxidation of the unsaturated fatty-acid chains in the alkyd binder then gives rise to hydroperoxides with uptake of atmospheric oxygen. Decomposition of these hydroperoxides results in the formation of peroxide and alkoxide radicals. These radicals initiate the polymerisation of the unsaturated molecules of the binding medium. Polymerisation occurs through radical termination reactions forming cross-links, causing gelling of the film, which is followed by drying and hardening. The number of cross-linked sites that are formed determines the film hardness. Cross-link formation is irreversible; hence when a paint layer has dried it cannot be easily removed.

Autoxidation is the direct reaction of molecular oxygen with organic compounds under relatively mild conditions. More specifically, autoxidation is described as the insertion of a molecule of oxygen into a C—H bond of a hydrocarbon chain to give an alkyl hydroperoxide. Autoxidation and metal-catalysed autoxidation has been extensively studied for numerous substrates under various reaction conditions. Generally, an induction time is observed after which the autoxidation reaction abruptly starts and rapidly attains a limiting, maximum oxidation rate. The reaction proceeds by a free-radical chain mechanism and can be described in terms of initiation, propagation and termination. Scheme 7.1 shows the generally accepted reaction steps.

Initiation reaction

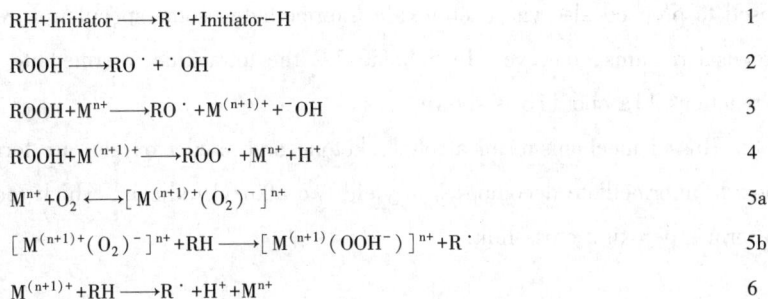

$$RH + Initiator \longrightarrow R^{\cdot} + Initiator\text{-}H \qquad\qquad 1$$

$$ROOH \longrightarrow RO^{\cdot} + {}^{\cdot}OH \qquad\qquad 2$$

$$ROOH + M^{n+} \longrightarrow RO^{\cdot} + M^{(n+1)+} + {}^{-}OH \qquad\qquad 3$$

$$ROOH + M^{(n+1)+} \longrightarrow ROO^{\cdot} + M^{n+} + H^{+} \qquad\qquad 4$$

$$M^{n+} + O_2 \longleftrightarrow [M^{(n+1)+}(O_2)^{-}]^{n+} \qquad\qquad 5a$$

$$[M^{(n+1)+}(O_2)^{-}]^{n+} + RH \longrightarrow [M^{(n+1)}(OOH^{-})]^{n+} + R^{\cdot} \qquad\qquad 5b$$

$$M^{(n+1)+} + RH \longrightarrow R^{\cdot} + H^{+} + M^{n+} \qquad\qquad 6$$

Propagation reaction

$$R \cdot + O_2 \longrightarrow ROO \cdot \qquad\qquad 7$$

$$ROO \cdot + RH \longrightarrow ROOH + R \cdot \qquad\qquad 8$$

$$RO \cdot + RH \longrightarrow ROH + R \cdot \qquad\qquad 9$$

Termination reaction

$$2 RO \cdot \longrightarrow ROOR \qquad\qquad 10$$

$$2 ROO \cdot \longrightarrow R = O + ROH + {}^1O^2 \qquad\qquad 11a$$

$$2 ROO \cdot \longrightarrow ROOR + O^2 \qquad\qquad 11b$$

$$ROO \cdot + M^{n+} \longrightarrow ROO^- + M^{(n+1)+} \qquad\qquad 12$$

$$R \cdot + M^{(n+1)+} \longrightarrow [R - M^{n+}]^{n+} \qquad\qquad 13$$

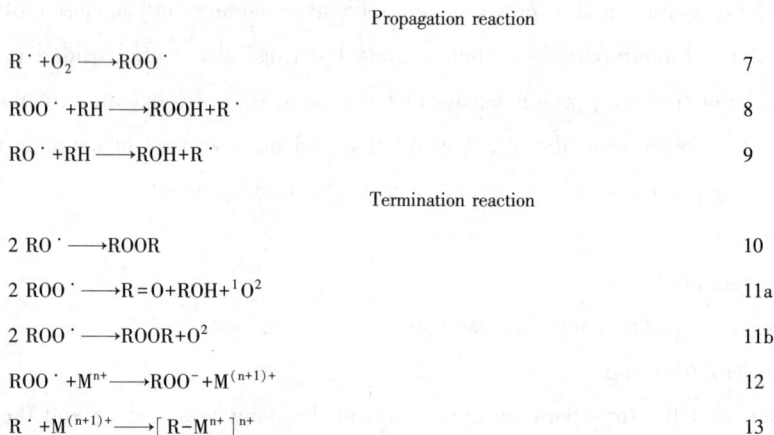

Scheme 7.1   Radical chain reactions as occuring in metal-catalysed autoxidation

Initiation can occur via several different pathways, either metal mediated or not. Steps 1 and 2 in Scheme 7.1 are often proposed as the initiation steps in non-metal catalysed reactions only peroxy radicals are of importance in chain propagation. Reaction 8 is relatively slow and thus it is the rate-determining step for the formation of hydroperoxides. Since the peroxy radical in reaction 8 is comparatively unreactive it has a strong preference for the abstraction of the most weakly bound hydrogen atom of the hydrocarbon substrate. One initiation event can set off the cycle between reactions 7 and 8, which is why these reactions are responsible for a large increase in hydroperoxide concentration during the first stages of autoxidation. Hydroperoxide formation will eventually be balanced by hydroperoxide decomposition, however, resulting in a steady state concentration of hydroperoxides in polar media. This is the main reason for autoxidation to attain a limiting rate. Alkoxide radicals can only be formed through metal-catalysed autoxidation, via reaction 3.

Termination reaction 10 is not a dominant reaction. Alkoxide radicals are much too reactive and hence their concentration is rather low. Under most autoxidation conditions, the only significant termination reaction is the reaction of two peroxyl radicals with each other(reactions 11a and 11b). Reaction 11a proceeds according to the Russell-mechanism, where a tetraoxide intermediate is formed which decomposes to yield non-radical products: an alcohol, ketone and singlet oxygen. The other bimolecular peroxyl radical termination reaction, reaction 11b, leads to a peroxide cross-link. This reaction is proposed to proceed also via a tetraoxide intermediate, some speculation regarding the validity of that proposal remains, however. In Scheme 7.2 the tetraoxide intermediate and the course of termination reactions 11a and 11b is shown.

In reaction 11a(the Russel mechanism) an alcohol, ketone and singlet oxygen are formed. In reaction 11, the tetraoxide intermediate decomposes to yield two alkoxyl radicals, which recombine in the solvent cage to form a peroxide cross-link.

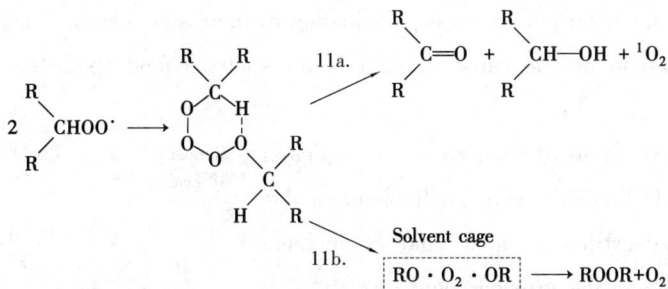

Scheme 7.2 Bimolecular peroxyl radical termination via a tetraoxide intermediate

Several aspects of alkyd paint autoxidation are typical for the alkyd system:

①Autoxidation occurs in an apolar environment.

②Oxidations take place at the fatty acid tail of the alkyd resin.

③Oxidation leads to cross-link formation.

Autoxidation drying starts when all solvent is evaporated. Consequently, autoxidation takes place in a mixture of pure alkyd resin and pigments, which is very likely a significantly apolar environment. This has some consequences for the autoxidation reactions, i.e. ionization processes will be suppressed and metal salts will not dissociate into ions. The most notable consequence for autoxidation of the alkyd system is that the Haber-Weiss reactions 3 and 4 do not occur in apolar media. Hydroperoxides are decomposed following metal-hydroperoxide complex formation, as shown in scheme 7.3. This has some far-reaching implications for metal catalysis, since reaction 4b in scheme 7.3 has been reported to be very slow for simple metal salts.

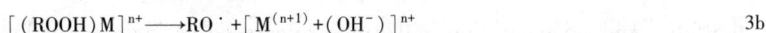

$$ROOH + M^{n+} \rightleftharpoons [(ROOH)M]^{n+} \qquad\qquad 3a$$

$$[(ROOH)M]^{n+} \longrightarrow RO^{\cdot} + [M^{(n+1)} + (OH^-)]^{n+} \qquad 3b$$

$$ROOH + M^{(n+1)+} \rightleftharpoons [(ROOH)M]^{(n+1)+} \qquad\qquad 4a$$

$$[(ROOH)M]^{(n+1)+} \longrightarrow ROO^{\cdot} + M^{n+} + H^+ \qquad 4b$$

Scheme 7.3 Metal-hydroperoxide complex formation in media of low polarity

Thus, reduction of the higher valence state metal is proposed not to be accomplished by hydroperoxides, but rather by easily oxidizable autoxidation products such as aldehydes and alcohols or directly by the substrate. Metal-hydroperoxide complex formation is also thought to account for the very sudden conversion of catalyst into inhibitor upon steadily increasing the metal concentration. In apolar media, metal salts will form a complex with a hydroperoxide as long as the hydroperoxide is available. Metal ions that are not coordinated can participate in the inhibiting reaction 12 in scheme 7.1, and thus sudden inhibition of autoxidation occurs if the metal concentration becomes higher than the hydroperoxide concentration.

The fatty acid tail of the alkyd resin is where autoxidation takes place. Fatty acids are important biomolecules, and are present in lipids as their triester with glycerol. Consequently, a considerable amount of research has been performed on elucidation of their autoxidation mechanism, since lipid autoxidation is known to be the cause of vital issues such as food spoilage, tissue injuries and degenerative diseases.

The fatty acids in an alkyd resin are polyunsaturated fatty acids, commonly linolenic acid, (α-linolenic acid = 9Z, 12Z, 15Z-octadecatrienoic acid and γ-linolenic acid = 6Z, 9Z, 12Z-octadecatrienoic acid) which is a major constituent of linseed oil, or linoleic acid(9Z, 12Z-octadecadienoic acid) which is a major constituent of, for example, sunflower oil and soya oil. The high susceptibility of non-conjugated polyunsaturated fatty acids(or lipids) for autoxidation comes from the presence of bis-allylic

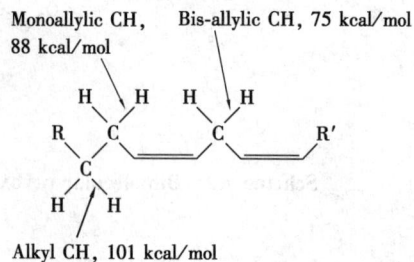

Monoallylic CH, 88 kcal/mol   Bis-allylic CH, 75 kcal/mol

Alkyl CH, 101 kcal/mol

Figure 7.2 Bond dissociation energies of the different CH bonds in fatty acids

hydrogen atoms, which have a relatively low bond dissociation energy of 75 kcal/mol(see Figure 7.2) and can therefore be easily abstracted, resulting in radical chain initiation and thus autoxidation. Abstraction of one of the bis-allylic hydrogen atoms results in the formation of a radical species. This radical species is stabilised by delocalisation due to the local pentadiene structure. Molecular oxygen reacts extremely rapid with this pentadienyl radical species to form a peroxy radical which has the double bonds dominantly conjugated, since this is the most stable structure. The peroxyl radical can then participate in a host of reactions, but in the early stages of autoxidation the dominant reaction will be to abstract a hydrogen atom from another lipid molecule to form a hydroperoxide and propagate the radical chain. Scheme 7.4 shows the initial autoxidation reactions for a fatty acid pentadiene substructure, forming a hydroperoxide. Figure 7.3 shows the total time course of fatty acid autoxidation: the fatty acid concentration will rapidly decrease as hydroperoxides are formed. The hydroperoxide concentration will go through a maximum when hydroperoxide formation is surpassed by hydroperoxide decomposition.

Hydroperoxide decomposition leads to further product formation, forming cross-linked (non-volatile) species and numerous other oxygen containing products such as alcohols, ketones, aldehydes and carboxylic acids.

7.2.2 Epoxy Coatings[8-9]

Epoxies are without question the most widely used coatings in the wastewater field. Epoxy coatings are generally made through the reaction of phenols with acetone or formaldehyde. Those reactants are then further reacted with epichlorohydrin. The resultant materials are diglycidyl ethers of what are called Bisphenol A epoxies, Bisphenol F epoxies, or phenolic novolac epoxies. These resins are then cross-linked via polymerization reactions with various curing agents or blends of curing agents. A basic discussion of the three main types of epoxy resins and the major categories of curing agents follows below.

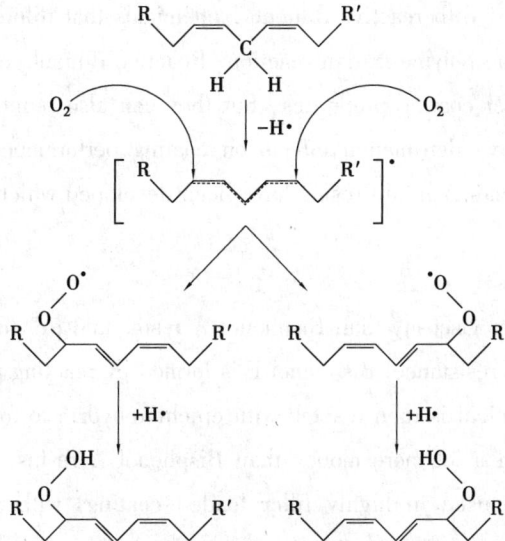

**Scheme 7.4   Initial hydroperoxide formation in the autoxidation of
the fatty acid chain of an alkyd resin binder unit**

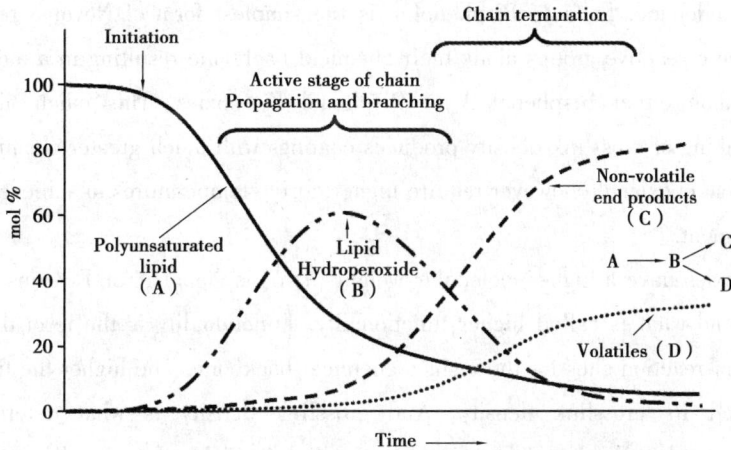

Figure 7.3   Time course of fatty acid(lipid)autoxidation

(1) Types of epoxy resins

1) Bisphenol A epoxies

These are the most commonly used resin for epoxy coatings. Bisphenol A resins are available in a large range of molecular weights. It is the reaction product of phenol and acetone. It is further reacted with epichlorohydrin. The resulting product is a thick liquid similar to honey in consistency. It is largely used for 100% solids coatings and flooring systems. Bisphenol A epoxy has good broad range chemical resistance, good physical properties, and is cured using a wide variety of curing agents at ambient temperatures. It is generally quite high in viscosity and this has historically limited its use in high filler loaded coatings. To reduce its viscosity for such uses, it has traditionally had solvents and diluents added to it. However, since the advent of strict VOC regulations, these

199

additions have been replaced with reactive diluents, chemicals that dilute or lower the viscosity of the resin while going into the polymerization reaction. Reactive diluents can be helpful in reducing viscosity and enhancing other coating properties, but they can also reduce the chemical resistance properties and otherwise have detrimental effects on coating performance. In more recent years, some lower viscosity Bisphenol A liquid resins have been developed which do not require the use of diluents or solvents.

2) Bisphenol F Epoxies

These resins have lower viscosity than Bisphenol A resins and provide much better strong acid and strong solvent chemical resistance. Bisphenol F is formed by reacting phenol with formaldehyde. The resulting phenolic chemical is then reacted with epichlorohydrin to form the Bisphenol F liquid resin. These resins also cost a lot more money than Bisphenol A resins. With lower viscosity, the Bisphenol F resins can be used in highly filler loaded coatings without the use of solvents or nonreactive or reactive diluents.

3) Novolac Epoxies

Like Bisphenol F Epoxies, Novolac epoxies are resins which are also formed via the reaction of phenol with formaldehyde. In fact, Bisphenol F is the simplest form of Novolac resins. However, Novolacs have more reactive groups along their chemical backbone resulting in a more highly cross-linked polymer than either Bisphenol A or Bisphenol F epoxies. This much higher degree of chemical crosslinking or crosslink density produces coatings with much greater chemical and thermal resistance. Novolac epoxies do however require higher curing temperatures to achieve their maximum property development.

Novolac epoxies have a higher molecular weight than Bisphenol A or F resins. This results in higher viscosity and what is called higher functionality. Functionality is the term that refers to the relative number of reaction sites for the resin's chemical backbone. The higher the functionality of a resin, the greater its crosslink density. And crosslink density is what determines chemical resistance. For example, Bisphenol F resins have a slightly higher functionality than Bisphenol A resins. See Table 7.2. This is largely why Bisphenol F resins have better resistance to a wider range of chemicals than Bisphenol A resins. Also, the chemical resistance of Bisphenol F resins is better due to its lower viscosity than Bisphenol A resins. This means that the use of fewer diluents or additives is required for viscosity reduction and those additives (as previously noted), also affect chemical resistance detrimentally. Due to this lower viscosity, Bisphenol F resins also remove the need for solvents from coatings. This reduces fire hazards and VOC.

Novolac epoxy resins have two distinct performance advantages over Bisphenol F resins. First, they have better chemical resistance due to their much higher functionality. See Table 7.2. This produces very high crosslink density. And secondly, the large quantity of aromatic ring structures increase the heat resistance of Novolac epoxies when compared to Bisphenol F resins.

Table 7.2　**Property of Bisphenol A/ Bisphenol F/ Novolac Resin**

Property	Epoxy Type		
	Bisphenol A	Bisphenol F	Novolac
Viscosity at 25 ℃	11 000~15 000 cps	2 500~5 000 cps	20 000~50 000 cps
Molecular	370	370	
Functionality	1.9	2.1	2.6~3.5

(2) Curing Agents for Epoxies

Due to ambient temperature curing requirements, almost all epoxy coatings must use amine based curing agents. While the selection of the epoxy resin generally establishes the limits on coating performance, the type of curing agent does affect coating properties in many ways. Within amine based curing agents, there are several classes that have different effects on coating performance and application properties. These include the following:

①Aliphatic polyamines

②Polyamine adducts

③Polyamide/Amidoamines

④Aromatic amines

⑤Ketimines

⑥Cycloaliphatic amines

1) Aliphatic polyamines

Aliphatic amines are multifunctional (meaning more than one reaction site per molecule). Aliphatic ethylene amines were the first curing agents or hardeners used in epoxy coatings. These hardeners provided high reactivity(fast cure)at ambient temperature and good solvent resistance due to high functionality. However, they are non-flexible and very phone to carbonation or blushing problems. As such, ethylene amines have largely been modified or blended with other hardeners to overcome these problems. Aliphatic amine cured epoxy coatings are used when strong chemical resistance is needed. They form tough coating films and the coatings have short pot lives and short cure times. These curing agents are very susceptible to what is called an "amine blush". This involves reaction between the aliphatic amines and moisture and carbon dioxide resulting in formation of an amine carbamate. This happens during cure of the coating. This blush results in a hazy discoloration of the coating and the formation of an oily film on top of the coating that can act as a bond breaker for subsequent top coats.

2) Polyamine adducts

Multifunctional(meaning more than one reaction site per molecule) aliphatic amines are partially reacted with epoxy resins to create amine adducts. The cured coating film(after the amine adduct is further reacted with epoxy resin in the coating) is similar to the aliphatic amine cured epoxy except the blushing problem is limited and reactivity is lower. The resulting coatings have longer pot lives and cure times as well. Generally, amine adduct cured epoxy coatings also have lower viscosities than aliphatic amine cured coatings.

3) Polyamide/Amidoamines

Polyamides and amidoamines (the low viscosity counterparts to polyamides) offer several advantages when compared to ethylene based aliphatic amines. This is due to the introduction of a fatty acid into the chemical backbone of the hardener. Polyamides and amidoamine curing agents are made by reacting aliphatic polyamines with fatty acids. The advantageous properties brought to coatings by polyamide and amidoamine curing agents include improved film flexibility, better wetting properties(and therefore adhesion), and good water resistance. Furthermore, these coatings are more tolerant of damp substrate conditions when applied without detriment to polymerization. As such, polyamide cured epoxy coatings do not develop amine blush problems. All of these advantages plus slower cure and longer pot lives come with lower functionality. Therefore, the chemical resistance of polyamide and amidoamine cured epoxy coatings (especially solvent and acid resistance) is greatly reduced when compared to amine cured epoxy coatings.

4) Aromatic amines

Aromatic Amines are based on the presence of an unsaturated ring of carbon atoms in the molecule. Common aromatic molecules include benzene and xylene. As such, aromatic amines include an amine functional group attached to a benzene ring structure. The presence of the benzene ring structure greatly enhances chemical resistance. Aromatic amines react quite strongly and therefore accelerators need to be added to speed up the rate of reactions. For many years the widest used aromatic amine was methylene dianiline(MDA). It was used in some of the most chemically resistance lining products ever provided. It also gave great heat resistance, a long pot life, and good flexibility. Its use, however, due to toxicity issues has now been outlawed. Alternative chemistries have now been developed by coating formulators to replace the MDA-like performance.

5) Ketimines

Ketimines are aliphatic amines that have been reacted with ketones to produce what is called "blocked amines". This means the amine is not able to cross-link with the epoxy resin until it is unblocked usually by the presence of water. The blocking provides longer pot lives to coatings and lower reactivity. Once the amine is unblocked, the coating film generally develops the same properties provided by the amine. The cure time for ketimines is very slow. And because the Ketone solvent must come out of the film as a volatile, there is a lot of opportunity for solvent entrapment related problems like pinholes.

6) Cycloaliphatic amines

This class of aliphatic amines is characterized by the presence of an amino group on the six carbon ring structure. These amines promote light stable coatings.

They also produce coatings with enhanced heat resistance when compared to other aliphatic amines. They also produce coatings with better mechanical properties when compared to aliphatic amine cured or polyamide cured epoxy coatings. They are slower to react than aliphatic amines, but are faster than polyamides or amidoamines. Chemical resistance is very good compared to polyamides or aliphatic amines.

7.3   Adhesives

Adhesives have been used by people for millennia. Some of the earliest, such as plant resins, starches and sugars, and concentrated proteins are still used for a variety of applications. In English, sticky substances used to join things together have been called by a variety of names. The word "adhesive" comes from a Latin root. "Cement", from an Old French root generally describes materials that become very hard after application. "Glue", from Old English, was usually used to describe protein extracted from animal parts but has a more general meaning now. "Lime", a word formerly used to describe any sticky substances (as in "bird lime"—plant gums used to trap birds by sticking them to the branches) survives now as a word for calcium hydroxide putty used to join stone and bind plaster. Even the word "mastic" from the Greek root meaning "to chew" was more generally used to describe a variety of sticky materials. In the past, all such substances were generally prepared by the user from locally available materials, only becoming commercial commodities in the last few hundred years[10].

The production of cheap and affective containers may have influenced the availability of proprietary adhesives and fillers just as they did in the case of art materials and drugs. There had already been a steady rise in the quantity of glass bottles produced, with accompanying decline in cost when the introduction of bottle-making machines in the mid-nineteenth century made them truly disposable. Other new and effective packaging systems included the collapsible metal tube, invented by John Rand in 1841, and the introduction of tinned sheet-iron "cans" or "tins" which were steadily improved throughout the nineteenth century. Such packaging improvements gradually eliminated the need for restorers to make their own adhesives. By the early twentieth-century proprietary adhesives and fillers were so common and convenient that some works on restoration from that period rely on them almost exclusively. More recently, conservators have found commercial packaging systems to be useful for the storage of adhesives formulated by the conservator.

The forces of attraction depend on very close proximity of molecules. A non-porous solid such as steel will attract itself if the mating surfaces are so perfectly polished that air is excluded and very close contact is made, but such close contact is not possible between most solid surfaces due to roughness, or because they are porous and consist mostly of voids (e.g., wood). This dictates the use of a liquid adhesive which can flow out onto a rough and void-filled surface "wetting" it intimately and serving as an intermediate between the solid surfaces. When the adhesive itself becomes solid by cooling (thermoplastics) chemical reaction (thermosets) or solvent loss (solution resins/adhesives), the adherends are firmly stuck together. The "pressure sensitive" adhesives are exceptions to this type of adhesion, being soft enough in solid form to conform very closely to surfaces. Pressure sensitive adhesives may also stick by a sort of micro suction cup effect generated by tiny voids on the surface of the adhesive. There is salt much that is not known about adhesion. What holds flies and gecko lizards on a ceiling is still subject to debate and competing theories.

In general, the better a substrate is "wet" by an adhesive, the better the bond will be because the degree of welling is itself dictated by the attraction generated between the substrate and adhesive ("specific adhesion"). A way to visualize this wetting is by the "contact angle" formed between the surface of a drop of liquid on a solid and the surface of the wet solid. A drop of water on a piece of wax for example "beads up" into a spherical shape having a large contact angle. A drop of water that can spread and flatten on an appropriate wettable surface has a small contact angle. Observation of this phenomenon can be a useful guide in judging the appropriateness of an adhesive, or of the condition of the adherend surface which may be too oily to be wet with a water based adhesive.

Given the ingenuity and imagination of human beings it is likely that almost any substance ancient or modem that could be construed as an adhesive has been tried as such. Many were obviously unsuitable, many have had brief vogues and fallen out of favor, and many have continued to be used under various service conditions. The primary literature should of course be consulted-especially for solutions to special problems but the following comments on the major categories of useful adhesives will serve as an introduction. For each category of adhesive brief summaries of the properties already discussed are given where they are notable.

7.3.1 Animal glues

Animal glues have great importance in the history of both woodworking and conservation and requires detailed discussion. They are some of the first adhesives used by man, and continued to be virtually the only adhesives employed in furniture construction until the relatively recent adoption of alternative adhesives for many applications in wood bonding. Animal glues have proven extremely durable under ideal circumstances and intact glue joins can be found that are centuries old. Optimum animal glue joins are stronger than the wood itself but even under less ideal circumstances offer adequate strength. They do not stain wood or impede the application of stains and coatings, are non-toxic and easily cleaned born areas where the adhesive is unwanted. Animal glues have been extensively used for the repair of antiquities and other art objects, both as primary adhesives, and as binders for filling materials and paints. Since they are water-soluble, restorations based on animal glues are relatively easy to reverse. Disadvantages of animal glues include limited working time at room temperatures, poor gap filling abilities, and bio-deterioration under some conditions. As water-soluble adhesives, animal glues are moisture sensitive. This makes them unsuitable for some applications. For furniture conservation, animal glues are recommended by the relative ease of reversibility. Because they are most likely to be the original adhesive in furniture glue joint, re-gluing an old join with animal glue is generally the best choice in terms of compatibility and strength.

Animal glues can be reversed with water, heat, or wet heat (steam). Glued joins can also be cracked apart by introducing alcohol into the joint. The alcohol desiccates the glue and follows small fractures into the join causing cohesive failure. This technique, sometimes called "dry cracking" can be used where the introduction of water or heat would be problematic.

Animal glues are derived from collagen ("colla", meaning "glue" and "gen", meaning "creator" in Greek) which is the protein present in skins, bones and connective tissue. Until fairly recently, the term "glue" meant only one thing, and that was animal glue. Terms such as "Flanders glue" and "Scotch glue," found in period literature refer to places where high quality glue was made. The bulk of animal glue is made from cattle hides and is called hide glue. Untanned skin-often tannery waste in the past is the preferred raw material for glue making. Contrary to popular understanding, hoofs are not used because protein is difficult to extract from this highly organized keratin structure. Other starting stock such as bones (bone glue) fish skins (fish glue) arid small-animal skins (rabbit-skin glue) are also used to produce distinctive adhesives. The material commonly called "gelatin" is an edible grade, made by filtering glue with activated charcoal in order to extract the color and objectionable smell.

Glue is made by using heat and water to hydrolyze the initial collagen, breaking and unstranding the collagen microfibrils into smaller extractable units (gelatin). The glue stock is first prepared by washing. Fats and oils are removed by saponification (soap formation) in a lime solution, and acid solutions are used to neutralize the lime as well as to remove unwanted mucus-type proteins. The glue stock is then steeped in water at controlled temperatures and a series of broths are drained off as they reach the desired protein concentration. Because the collagen will continue to break down by hydrolysis as the temperature rises and the steeping time increases, a series of extracts of progressively lower molecular weight proteins can be cooked from the glue-stock at progressively higher temperatures until it is exhausted. The highest quality glue comes from the first draining and concentration of the glue-broth. In modem practice, the extracts are dehydrated by low temperature boiling in a vacuum so that they will not be further damaged by high heat. The concentrated broth is cooled until it gels and air-dried to the final product.

Animal glues made from mammalian skins possess unique properties that have made them useful in a wide range of applications. A small amount of glue can tie up large quantities of water into a gel structure. The flexible and rubbery gels consist of expanded open networks of protein polymer chains which hold water in the structure by hydrogen bonding and other forces. This gel state is temperature dependent. Animal glue is applied hot and fully fluid in a water solution that can readily wet a polar substrate such as wood. Setting occurs when the glue gels upon cooling into a rubbery elastomeric state. The gel then dehydrates and contracts until it is a hard and tough solid.

Starches are polymers made up of long chains of simple sugars(polysaccharides). Starches are present in plants in the form of complex granules with no adhesive characteristics. These granules are essentially food storage capsules. If starch granules from any plant source(rice, potato, wheat), are cooked they swell and burst, yielding a water-dispersed colloidal "paste" with good adhesive qualities. Starch adhesives are very polar and wet equally polar cellulosic structures such as paper and wood well. Cellulose itself is a polysaccharide. Starch paste has been used most extensively in paper and ethnographic conservation. In Japan however, where complex worked joins often do the main job of holding wooden structures together starches have been used for woodworking. Starch and plant gum mixtures have also been used for the adhesive backing of textiles. Starch and sugar "pastas" have been used for the traditional lining of easel paintings.

Dextrins are derived from starches and are also polysaccharides. They are made by hydrolyzing starches to create shorter chain lengths, then re-polymerizing the low molecular weight fragments. Extensively used in bonding paper due to quick tack time, they have seen little use in woodworking or other applications requiring high strength. The viscous, amber colored "mucilage" type adhesives sold in office supply stores are dextrin based.

Plant gums that are completely or partially soluble in water are also polysaccharides. There are a large number of plant sources. One of the most common is gum Arabic from members of the Acacia genus. Gum Arabic is the usual binder in water-color paints, and was also used extensively as a binder in restoration fills.

7.3.2　Natural resins

Natural resins are the naturally occurring, water-insoluble hard organics. They come from a variety of plant and animal sources: resins from trees derive born angiosperms and gymnosperms, both living and extinct(amber and some varieties of copal). They include damar, mastic, sandarac, resin and copal resins(the so-called "fossil" resins). Also from trees are the various oleo-resins with volatile components called "turpentines" and "balsams" (Strasbourg turpentine, Venice turpentine, Bordeaux turpentine and Canada balsam). Shellac is exuded by scale insects. A component of Mexican lacquer is also extracted from scale insects.

Natural resins generally consist of repeat units of isoprene and are classified as monoterpenes isoprene units includes oil of turpentine, lavender oil sesquiterpenes(3 units: includes shellac) diterpenes(4 units: includes rosin, sandarac, copals) triterpenes (6 units: includes mastic, dammar) and long chain polyterpenes(n units: includes natural rubber). As with most natural materials, detailed basic books on occurrence, trade, nomenclature and use date from the end of the last century and the beginning of this one.

Natural resins dissolved in oils were generally called "varnish", and show up in fill recipes fairly frequently. Resins dissolved in solvent are more usual as adhesives. Resins may also be used in fill and adhesive recipes as relatively minor additives meant to modify the properties of the bulk adhesive. When added to animal glue recipes for example, as in gilder's composition and "diamond cement", natural resins and varnishes increase water resistance and help to mitigate shrinkage. In the thermoplastic "mastics," natural resins are the major constituents. A typical historic example is the filling mastic for flaws in marble consisting of "yellow wax, rosin, burgundy pitch and a little sulfur and plaster" colored to match with suitable pigments.

A few of the natural resins should be described a little more fully due to common and widespread use in a variety of cultures.

①Resin(colophony): Rosin is the solid component of raw pine resin after the volatile "spirits of turpentine" have been distilled off. An alternate name for rosin is colophony(from the Greek words for glue and sound) which refers to its use on bows for stringed instruments. It was a cheap natural resin during the past few hundred years, and was used extensively for low-quality varnishes, for adhesives and for filling materials. Cutler's cement for filling the hollow handles of silver knives and holding the blades in place was commonly made from colophony, sulfur and any convenient bulking agent.

②Pitch and tar: These are inexact terms which may refer to a variety of substances including various naturally-occurring minerals(also called bitumen and asphalt), to the products of the pyrolysis or destructive distillation of both hardwoods and softwoods, or to the less volatile compounds present in fossil coal and crude oil. All of these materials are dark-brown to black in color, sticky, and cheap in comparison to other natural resins, even including rosin. They have all been used since antiquity(with the exception of refined petroleum tar) for caulking seams, halting weapons, water-proofing textiles and cordage(sailors were called jack tars), and generally filling gaps and sticking things together. Bitumen fills have been found as ancient repairs to ceramics. All of these materials can be and have been mixed with a variety of bulking agents to form filling compounds.

③Shellac: Shellac is the resin exuded by the scale insects, primarily *Kerria lacca*(Kerr). It has been used as a coating, adhesive and binder in various bulked compositions, including those from which photograph cases and phonograph records were made. Dissolved in alcohol, it has been a popular adhesive for adhering ceramics, stone, and other materials, and sticks of shellac, sometimes softened by the addition of wax or some other softer resin, were used as a thermoplastic fill material. Solid shellac sticks called "burn in sticks" have been extensively employed by furniture restorers to fill small damages. Shellac was a popular adhesive for the repair of antiquities in particular. One method of use was to apply shellac in alcohol to the break edges, then ignite the alcohol so as to bum off the solvent and heat the shellac. The parts were then immediately put together. Bonding is virtually instantaneous and so the assembly is rapid. Large-scale export of shellac resin from India increased rapidly after 1870 indicating that shellac as a common and cheap "plastic"(as opposed to an adhesive) may well post-date the 1870s.

7.3.3 Synthetic resins

Man made resins or "plastics" have been and are currently used both as adhesives and as coatings. Virtually all of the tremendous variety now available has had some use, however brief, in adhesives. Synthetic resins are used as adhesives in two distinct forms; un-modified straight-chain resins dissolved in solvents, and extensively modified emulsions which are initially dispersed in water.

Both natural and synthetic resins dissolved in various solvents have seen extensive use as coatings, but have also been used by conservators as consolidants for degraded wood and poorly bound or adhered coatings. The use of resin consolidants is a complex topic better addressed elsewhere. Many solvent-release, solution-type adhesives are used in restoration and conservation.

(1) Cellulose nitrate and cellulose acetate

Cellulose nitrate resins, developed in the mid-nineteenth century were the first synthetic polymers, and were put into use as adhesives at an early date. Cellulose nitrate dissolved in solvent was termed "collodion" and as such was sometimes used as a binder. By itself, collodion was excessively brittle and so was often plasticized with other materials. One "elastic collodion" cement was made with "gun cotton" (cellulose nitrate) dissolved in ether and alcohol with the addition of Venice turpentine and castor oil as plasticizers. Celluloid was plasticized with camphor, and this too was sometimes dissolved in solvent to make adhesives.

Cellulose acetate, first synthesized in 1894, has also been used for stiffening textiles used on aircraft and model aircraft ("dope"). Solvent type adhesives based on cellulose-nitrate, cellulose-acetate or both, (H.M.G., Duco, UHW™ etc.) have seen extensive use by hobbyists as well as in conservation. The long term stability of cellulose nitrate has been debated in the literature, with some authorities believing that it offers acceptable permanence on some substrates, and others condemning it as chemically unstable.

(2) Rubber and gutta-percha

Natural rubber and gutta-percha are both derived from hydro-colloid plant latexes. They have identical chemical makeup (polyisoprene) but are stereo-isomers (right and left handed molecules) of each other, with rubber being cis, and gutta-percha and a related natural latex called balata being trans. Natural rubber, also called "India rubber", is elastic and rubbery while gutta-percha is only soft while warm, becoming rigid and flexible at room temperature. Both natural rubber and gutta-percha were used alone and in mixtures with other materials as adhesives. In 1922, rubber was synthesized by Hermann Staudinger. Natural and synthetic rubbers as well as gutta-percha are prone to degradation through oxidation and de-polymerization. Rubber first becomes sticky as it deteriorates, then eventually becomes hard and crumbly, while gutta-percha degrades by falling into chunks, and eventually powder.

(3) **Epoxies**

Epoxy or epoxide resins are a group of reactive compounds that are characterized by the presence of the oxirane group[11].

They are capable of reacting with suitable hardeners to form cross-linked matrices of great strength and with excellent adhesion to a wide range of substrates. This makes them ideally suited to adhesive applications in which high strength under adverse conditions is a prerequisite. Their unique characteristics include negligible shrinkage during cure, an open time equal to the usable life, excellent chemical resistance, ability to bond nonporous substrates, and great versatility. Although they were hailed as wonder products when first introduced, it has now been accepted that they will not do everything. They have, however, clearly established niches, especially in high-technology applications, and have shown steady growth, generally ahead of the industry average. Sales of epoxy resins in Europe, for example, totaled 101 000 metric tons in 1980, 150 000 metric tons in 1985, and 205 000 metric tons in 1990.

Although work on epoxy resins started in the mid-1920s, the first commercially useful epoxy resins appeared during World War II. These were based on the diglycidyl ether of bisphenol A (DGEBA), and today these resins, in a range of molecular weights, constitute the majority of all epoxy resins used. By contrast, however, hardeners come in a variety of shapes and sizes, including amines and amides, mercaptans, anhydrides, and Lewis acids and bases. Choice of hardener depends on the application requirements, and the wide range of hardeners available increases the versatility of adhesives based on epoxy resins.

Epichlorhydrin is capable of reacting with hydroxyl groups, with the elimination of hydrochloric acid. The most widely used epoxy resins are the family of products produced by the reaction between epichlorhydrin and bisphenol A.

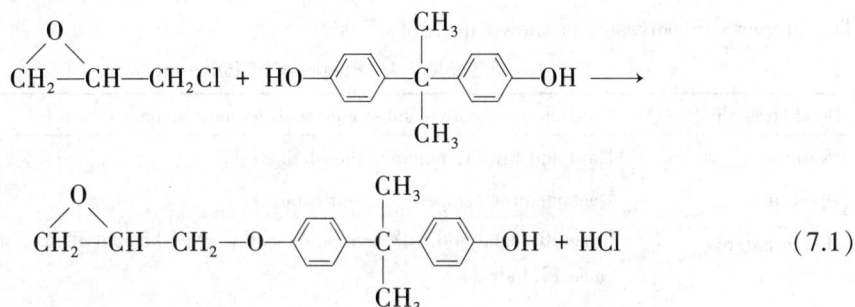

$$
\underset{O}{CH_2-CH-CH_2Cl} + HO-\!\!\!\left\langle \bigcirc \right\rangle\!\!\!-\underset{CH_3}{\overset{CH_3}{C}}-\!\!\!\left\langle \bigcirc \right\rangle\!\!\!-OH \longrightarrow
$$

$$
\underset{O}{CH_2-CH-CH_2-O}-\!\!\!\left\langle \bigcirc \right\rangle\!\!\!-\underset{CH_3}{\overset{CH_3}{C}}-\!\!\!\left\langle \bigcirc \right\rangle\!\!\!-OH + HCl \tag{7.1}
$$

Epoxy resin adhesives are used mainly in niche applications rather than as general-purpose adhesives. Due to the high strengths that can be achieved and the relatively high costs, they are generally used in structural applications in both concrete and metal bonding. Their good electrical properties allied to low shrinkage and good durability suit them for potting and encapsulating. Low shrinkage and good gap filling make epoxies ideal for applications where clamping is difficult, while the fact that both components are generally liquid up to the moment of cure means that they can be used where applications constraints require long open or assembly times. Conversely, systems with very short cure times are perfect for consumer applications. Good adhesion to nonporous surfaces allows them to be used in demanding situations. They find major outlets in the construction, automotive, and electronics industries. The property of epoxies resin is shown in Table 7.3.

Table 7.3 **Property of Epoxies Resin**

Bond strength	Excellent adhesive and cohesive strength
Hardness	Hard and brittle to rubbery depending on formulation, creep very low even for soft varieties
Open time	Highly variable depending on formulation
Health hazards	Uncured resins and hardeners are toxic, cured films are not
Color	White to dark amber. All will darken with age but this affect is very slight for some varieties. Optical saturation of substrates is high
Water resistance	Epoxies are waterproof
Reversibility	Irreversible on porous substrates, good on hard and non-porous substrates such as porcelain and glass(generally requires use of chlorinated solvents).

(4) Polyesters

Polyester resins, available since 1946 are condensation polymers of variable structure. They may be thermoplastic linear chains(Melinex, Mylar), unsaturated(UP) thermosetting polymers such as are used in glass reinforced resin(GRP)popularly known as "fiberglass", or as the class of coatings known as alkyds. As adhesives, polyesters have been used by stone conservators because of good gap-filling when bulked, fairly rapid cure, and good strength in high-load situations. The stone adhesives manufactured by Tiranti (UK) and Akemi (Germany) are examples. Polyesters were extensively used as filling materials, but since they yellow more than epoxies, shrink more, and are more objectionable to work with, epoxy resins have largely taken over in conservation applications. The property of polyester is shown in Table 7.4.

Table 7.4 **Property of Polyester**

Bond strength	Good on semi-porous substrates such as most stone
Hardness	Hard and brittle, tougher when bulked
Open time	Dependent on temperature and catalyst
Health hazards	Generally catalyzed with styrene or methyl-ethyl ketone(MEK)both are toxic. MEK has caused blindness
Color	Clear
Water resistance	Waterproof
Chemical stability	Prone to photo-oxidation, yellows with time
Reversibility	As a cross-linked polymer, similar to epoxy

(5) Polyurethanes

Polyurethane resins are built from units of polyesters or polyethers. First developed in 1943, they are now extremely variable in properties, and may be thermoplastic or thermosetting. In addition to the well-known finishes, they are increasingly important as molding and casting rubbers. Polyurethane adhesives are recently available on the market(Titebond Polyurethane, Gorilla Glue) and are being used increasingly by woodworkers and do-it-yourself repairers. The property of polyurethane is shown in Table 7.5.

Table 7.5   Property of Polyurethane Resin

Bond strength	Excellent, requires moisture to cure(wood must be damp)
Hardness	Hard and creep-resistant.
Open time	Long(fifteen to twenty minutes)
Health hazards	May provoke allergic reactions, stains and bonds to skin
Water resistance	Excellent
Chemical stability	Degraded primarily photo-oxidation, so not a problem with adhesives
Reversibility	As a cross-linked resin, similar to epoxies

(6) Cyanoacrylates

These adhesives are popularly called "superglues" after an early trade name. The first of these adhesives(Eastman 910^3W) was introduced in 1958 and available brands and physical properties have increased since. Unlike the thermosetting resins described so far, no separate hardener is required for curing. The fluid adhesive polymerizes very rapidly in contact with a weak base, or from the hydroxyl groups present in the thin film of water that is present or on most substrates. Cyanoacrylates have the chief advantage of very rapid cure which eliminates the need for any clamping but hand pressure. They will stick to a wide variety of substrates(especially skin)and so are useful in bonding dissimilar materials. Initially, cyanoacrylates had virtually no gap-filling abilities and would not cure in any but very thin glue lines. New varieties are available which have good gap-filling characteristics, and they have found favor with woodworkers for quick repairs of small cracks and losses. Cyanoacrylates may have poor long term chemical stability, but no thorough conservation studies have so far been done. Adhesive joins may become weaker and prone to failure as they age. These fears have limited their use in conservation. The property of cyanoacrylates is shown in Table 7.6.

Table 7.6   Property of Cyanoacrylates

Bond strength	Excellent but may decrease with time. Relatively poor peel strength
Hardness	Hard, machinable, low creep
Open time	Measured in seconds, shorter on alkaline substrates, longer on acidic substrates(wood)
Health hazards	Bonds to skin(used medically for closing wounds)
Color	Clear
Chemical stability	Short shelf life, possibly poor long term stability when cured
Reversibility	Poor on porous substrates, good on non-porous substrates(swellable but not soluble in acetone)

(7) Silicone rubbers[12]

Silicone resins have low shear properties but excellent peel strength and heat resistance. Silicone adhesives are available as either solvent solutions for pressure sensitive adhesives or as one or two part liquids.

Silicone adhesives are generally supplied as solvent solutions for pressure-sensitive application. These systems cure via a condensation and radical polymerization process. The resulting adhesives are very tacky and exhibit only moderate peel strengths as a result of their very poor cohesive strength. The adhesive reaches maximum physical properties after being cured at elevated temperature with an organic peroxide catalyst. A lesser degree of adhesion can also be developed at room temperature.

Silicone adhesives retain their adhesive qualities over a wide temperature range, and after extended exposure to elevated temperature. They are very tacky materials that bond to a wide variety of substrates. Because silicones have a relatively low surface energy, they bond well to many low surface energy plastics such as polyethylene and fluorocarbons. Pressure sensitive silicone adhesives are often used in the form of pressure sensitive tapes. A large application is pressure sensitive tape used in the electronics industry for various applications on printed circuitry. They are also often used as adhesives on silicone rubber backing for gasketing on ovens and other high temperature apparatus. Room temperature vulcanizing(RTV)silicone-rubber adhesives and sealants form flexible bonds with high peel strength to many substrates. These adhesives are one-component pastes that cure by reacting with moisture in the air. Because of this unique curing mechanism, adhesive bond lines should not overlap by more than 1 inch. RTV silicone materials cure at room temperature in about 24 hrs. With most RTV silicone formulations, acetic acid is released during cure. Consequently corrosion of metals, such as copper and brass, in the bonding area may be a problem. For corrosive substrates, certain RTV silicone formulations are available that cure by liberating methanol rather than acetic acid.

(8) Hot melt adhesives[11]

Hot melts are a widely used class of adhesives that are used for many applications but are rarely used for structural bonding, as seldom are they able to match the tensile strengths of other adhesive classes. Their primary uses are in packaging and in wood for edge veneering and veneer splicing. There are important reasons for employing hot-melt adhesive systems, such as:

①Ease of Application via high-speed equipment

②Formation of strong, permanent, and durable bonds within a few seconds of application

③No environmental hazard and minimal wastage because of 100% solid systems

④Ease of handling

⑤Absence of highly volatile or fiammable ingredients

⑥Excellent adhesion

⑦Wide formulation possibilities to suit individual requirements (e. g., color, viscosity, application temperature, and performance characteristics)

⑧Cost-effectiveness

Hot melts are 100% solid thermoplastic materials that are supplied in pellet, slug, block, or irregular-shaped chip form. They require heating via appropriate application equipment, which usually is fairly sophisticated in order to control the required temperature and coverage rate. Upon application, the heat source is removed and the thermoplastics set immediately (within a few seconds). Hot melts are thus well suited to high-speed continuous-bonding operations.

Here, ethylene-vinylacetate(EVA)hot melt adhesives will be mainly introduced.

Edge veneering requires use of a hot-melt adhesive that is relatively high in viscosity at application temperatures(usually around 200 ℃). The reasons for this are as follows:

①The adhesive must have sufficient body to prevent flowing from vertical surfaces after application.

②It must not penetrate the substrate surface too deeply, causing glue starvation.

③It must have easy spreading and excellent wetting characteristics.

Viscosities of these hot melts are on the order of 50 000 to 60 000 m Pa.s at 200 ℃. Viscosity is achieved through the correct selection of ethylene-vinylacetate copolymer grades, coupled with the quantity and type of reinforcing filler that is added to the system. The ball and ring softening point is an early indication of the degree of heat resistance of a particular hot melt. The softening point is influenced by the combination of ingredients, but to a large extent by the grade and quantity of EVA copolymer and tackifying resin contained in the system. Using a 5. 1g lead ball, the average softening points are between 90 to 105 ℃.

For optimum adhesion, the wetting characteristics (of the hot melt to substrates during application)are vital. Proper wetting is related to viscosity but is again largely influenced by resin selection and quantity. Stability of the adhesive is another important consideration. During prolonged periods at elevated temperature while contained in the hot-melt applicator, the hot melt must resist oxidation and thermal breakdown of components. This often leads to discoloration, charring, and inferior bonds. As a result of charred material, nozzle blockages can also be encountered.

EVA hot melts consist basically of the following:

①EVA copolymer

②Tackifying and adhesion-promoting resins (e. g., hydrocarbon, rosin esters, coumarone-indene, terpene resins)

③Fillers, usually barium sulfate(barytes)or calcium carbonate(whiting)

④Antioxidants

Because of their relatively high melt viscosities, the EVA hot melts need special manufacturing equipment. For example, a Z-blade mixer such as a Baker Perkins or Winkworth with oil-heated jacketing is required. Mix temperatures are kept as low as possible(about 110 ℃) to keep bulk thick. The high-viscosity kneading action ensures rapid dissolution of EVA copolymer and resin.

Fillers are easily dispersed and a homogeneous mix is achieved rapidly with this type of agitation. Upon completion, the molten product is extruded into ropes approximately 6 mm in diameter, which are cooled through a chilled water trough and then granulated into pellet form. Alternatively, hot-melt slugs are supplied where application equipment utilizes this form. It is essential to ensure that any residual moisture picked up during the cooling process is eliminated via an air-drying cyclone before packing.

Collection of Exercises

1. Please explain the meaning of the following words: adhesive, adhesion and cohesive?

2. How many types of adhesives can you list? Give some examples.

3. What are main resins of adhesives? And their advantages and disadvantages?

4. Please explain the importance of contact angle of a surface to adhesion.

5. Please describe the process of adhesion.

6. Why does epoxy adhesive can bond tightly with metals?

7. What is the mechanism of polyurethane adhesives' reaction? What are reasons of urethane adhesives that make them good adhesives?

8. What are main application fields of adhesives?

9. How does a coating work? In other words, what are the processes of film-forming?

10. What are main basic ingredients of coatings? And explain the function of resins in coatings.

11. Name the main types of waterborne coating for inner walls of houses.

12. What are main categories of pigments? Explain the differences between pigment and dye in coating use.

13. What do environment-friendly coatings include? Explain why they are environment-friendly?

14. Which kind of coatings is most suitable for exterior application for a car? And why?
    (1) Acrylic coatings;
    (2) Alkyd coatings;
    (3) Epoxy coatings.

REFERENCES

［1］http://www.CEPE.org，网站资料.

［2］http://www.paint.org，网站资料.

［3］http://www.corrosion-club.com/historycoatings.htm，网站资料.

［4］http://home.nycap.rr.com/useless/paint/index.html，网站资料.

［5］R. van Gorkum, E. Bouwman, The oxidative drying of alkyd paint catalysed by metal complexes, Coordination Chemistry Reviews, 249: 1709-1728, 2005.

［6］Charles A. Harper, Modern plastics handbook, McGraw-Hill Professional, 2000.

［7］Remy van Gorkum, Manganese Complexes as Drying Catalysts for Alkyd Paints, Leiden University, Faculty of Mathematics & Natural Sciences, Dept. of Chemistry, 2005.

［8］http://www.hrsd.com/pdf/Coatings% 20Manual/2011/APPENDIX% 20C.pdf, Basics on Coatings Chemistry, Coatings Manual, 2011.

［9］John D. Durig, Comparisons of Epoxy Technology for Protective Coatings and Linings in Wastewater Facilities, the Journal of Protective Coatings & Linings, Technology Publishing Company, 2000.

［10］Jonathan Thornton, Adhesives and Adhesion, Buffalo State College, 2005.

［11］A. Pizzi, K. L. Mittal, Handbook of Adhesive Technology, Second Edition, Marcel Dekker, Inc., USA, 2003.

［12］Edward M. Petrie, Handbook of Adhesives and Sealants, Second Edition, McGraw-Hill Professional, 2007.

CHAPTER **8**

FUNCTIONAL POLYMERS

8.1 Introduction

Chemical reactions of polymers have received much attention during the last two decades. Many fundamentally and industrially important reactive polymers and functional polymers are prepared by the reactions of linear or cross-linked polymers and by the introduction of reactive, catalytically active, or other groups onto polymer chains. According to the International Union of Pure and Applied Chemistry(IUPAC) Commission on Macromolecular Nomenclature, a functional polymer is a polymer that exhibits specified chemical reactivity or has specified physical, biological, pharmacological, or other uses that depend on specific chemical groups[1]. The term "functional polymer" has two meanings:①a polymer bearing functional groups (such as hydroxy, carboxy, or amino groups) that make the polymer reactive and ②a polymer performing a specific function for which it is produced and used. The function in the latter case may be either a chemical function such as a specific reactivity or a physical function like electric conductivity. Polymers bearing reactive functional groups are usually regarded as polymers capable of undergoing chemical reactions.

Functional polymer material is a research area which has the fastest rate of development in polymer science field and has the highest degree of interdisciplinarity comparing to other areas of science. It is an academic discipline that builds on the foundation of polymer chemistry, polymer physics and other related subjects. The academic discipline also strongly linked to physics, medical science, even biology. Functional polymer material is a kind of polymer and their composites material which are able to transmission, transform or store the substances, energy and information, sometimes also called fine polymer or special polymer (including high performance polymer). Functional polymer material began to develop rapidly in the end of 1960s as a new kind of polymer material. It is rich in content, has a great variety, develops rapidly and became known as an essential critical material in new technical revolution.

Based on original mechanical properties of synthetic or natural polymer, polymers which have many kinds of specific function (such as chemical activity, biocompatibility, pharmacological properties, select classification and so on) other than the traditional properties have been made. Generally speaking, there are many groups in the backbone and side chain that show certain function. The display of the function are usually very complex, because it is not only depend on the primary structure like the chemical structure of polymer chain, the sequence distribution of structure unit, molecular weight and distribution, branching and stereochemical structure, but also determined by conformation of polymer chain, the high-level structure when polymer chain is gathering, etc.

In this chapter, attention is focused on a several kinds of functional polymers. These include liquid crystal polymers, conductive polymers and ion exchange resins.

8.2   Liquid crystal polymers

Ordinarily a crystalline solid melts sharply at a single, well-defined temperature to produce a liquid phase that is amorphous and isotropic. A different behavior is exhibited by a class of organic compounds known as liquid crystals. The oldest examples are cholesterol derivatives, e. g., cholesteryl benzoate. This substance, for instance, does not have a sharp transition to amorphous liquid at 145.5 ℃, but changes to a cloudy liquid, which becomes clear and isotropic only at 178.5 ℃. This cloudy intermediate state that possesses an ordered structure with some resemblance to a crystalline solid, while still in the liquid state, is called a mesophase or mesomorphic phase from the Greek "mesos", meaning in between or intermediate[2].

Liquid crystal materials can thus be defined as those which are characterized by the appearance of mesophases between the crystalline solid and isotropic liquid phases. Many organic compounds with this property have been synthesized. The temperature range of stability of a mesophase may be anything up to 150 ℃.

Liquid crystals can be divided into two main classes; those forming liquid crystalline phases in the melt are called thermotropic and those forming liquid crystalline phase in solution are referred to as lyotropic. Depending on the type of molecular order in the mesophase these classification can be broken down further into three main categories: smectic, nematic, and cholesteric.(Figure 8.1)

Smectic liquid crystals consist of elongated molecules that line up with the long axes of the molecules aligned in one direction and the ends of the molecules lying on parallel planes to produce a type of layered structure (Figure 8.1(a)) like that in a layered box of cigars. A layer of smectic molecules is just one molecule (longitudinally) thick. The molecules in nematic liquids are similar

217

in shape to smectic molecules and also point in one direction, but unlike the latter they do not line up with the ends lying on parallel planes (Figure 8.1(b)).

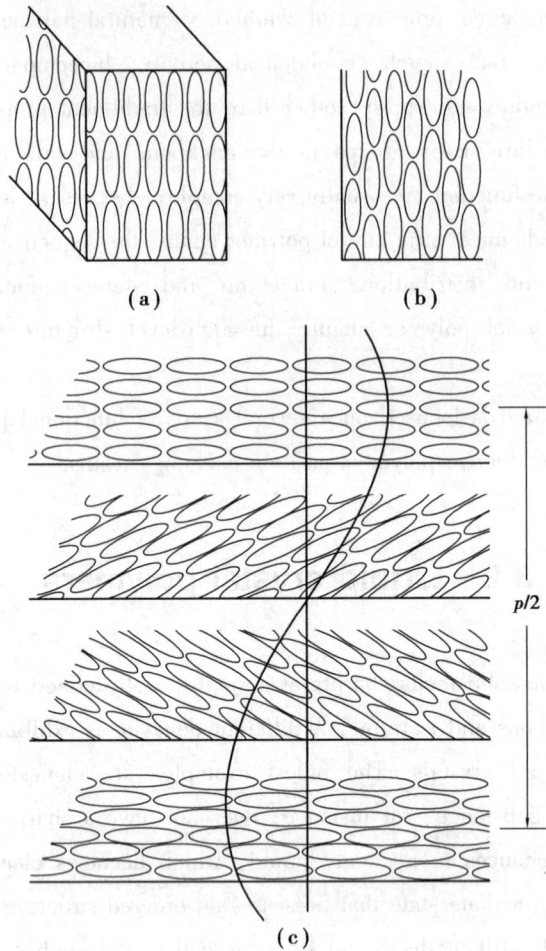

(a) (b)

(c)

Figure 8.1   Liquid crystals

(a) Smectic crystals: the ends of the molecules are on a plane.

(b) Nematic crystals: the ends of the molecules do not match.

(c) Cholesteric crystals: the molecules in each layer are arranged in

a manner similar to nematic crystals, but the angle changes from plane

to plane of the molecules, forming a helix of pitch length p.

Cholesteric liquid crystals consist of long flat molecules that line up in the same manner as nematic molecules with molecular long axes parallel to each other in a plane. However, the molecules in one plane are slightly displaced (due to side groups of molecules) with respect to those in the neighboring planes to form a helical pattern, as indicated Figure 8.1(c). Since the cholesteric structure is basically a derivative of the nematic, a transition from the cholesteric structure to the nematic structure can be effected by using a magnetic or electric field to unwind the helical configuration of the former.

Liquid crystals have interesting electro-optical properties. When subjected to small electric fields, reorientation and alignment of the liquid crystal molecules takes place, which produces striking optical effects because light travels more slowly along the axes of the molecules than across them. This has led to their use in optical display devices for electronic instruments such as digital voltmeters, desk calculators, clocks, and watches. Nematic liquid crystals are most commonly used in these applications. Cholesteric materials are added to provide memory effects.

Molecules that have a tendency to form liquid crystalline phase usually have either rigid, long rod-like shapes with a high length to breadth (aspect) ratio, or disc-shaped molecular structures. These rigid groups, referred to as mesogens, may be chemically composed of a central core comprising aromatic or cycloaliphatic units joined by rigid links, and having either polar or flexible alkyl and alkoxy terminal groups.

Polymers exhibiting liquid crystalline properties can be constructed from these mesogens in three different ways: ① incorporation into chain-like structures by linking them together through both terminal units to form main-chain LCPs; ② attachment through one terminal unit to a polymer backbone to produce a side-chain comb-branch structure; and ③ a combination of both main-and side-chain structures. LCPs typically show either smectic or nematic liquid crystal behavior. A schematic diagram of the two main phase types is shown in Figure 8.2.

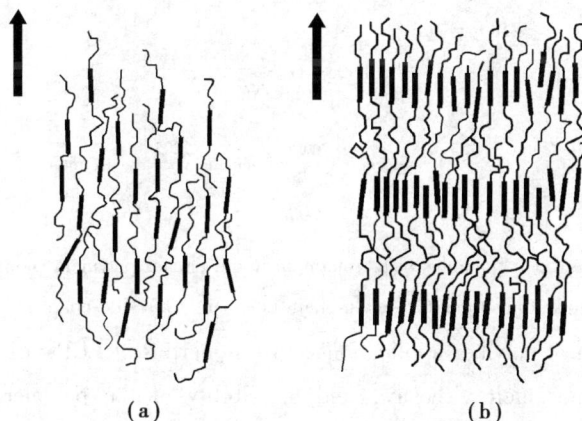

(a)   (b)

Figure 8.2   Schematic representation of (a) nematic phase and
(b) smectic phase for main-chain liquid crystalline polymers, showing the director
as the arrow. The relative ordering is the same for side-chain-polymer liquid crystals

The liquid crystalline phases in polymeric materials are sometimes difficult to identify unequivocally. However, several techniques can be used that provide information on the nature of the molecular organization within the phase, and if used in a complimentary fashion these can provide reliable information on the state of order of the mesogenic groups.

While differential scanning calorimetry is widely used as a means of detecting the temperatures of thermotropic mesophase transitions, the phases can be identified by observing the characteristic textures developed in thin layers of the polymer when viewed through a microscope using a linearly polarized light source. X-ray diffraction can be used to characterize the mesophases and to provide reliable information on the number of phases present.

The type of phase formed in a polymer liquid crystal can also be identified often by examining the manner in which it mixes with a small molecule mesogen of known mesophase type. If these textures are the same, then a mixed liquid crystal phase is formed with no observable transition between the two types of molecule.

In some polymer crystals, moreover, several mesophases can be identified. In main-chain LCPs there is usually a transition from the crystal to a mesophase, while in more amorphous systems where a glass transition (T_g) is present, the mesophase may appear after this transition has occurred. In thermotropic system having multiphase transitions, the increase in temperature leads to changes from the most ordered to the least ordered states, i.e., crystal/smectic/nematic/isotropic (see Figure 8.3).

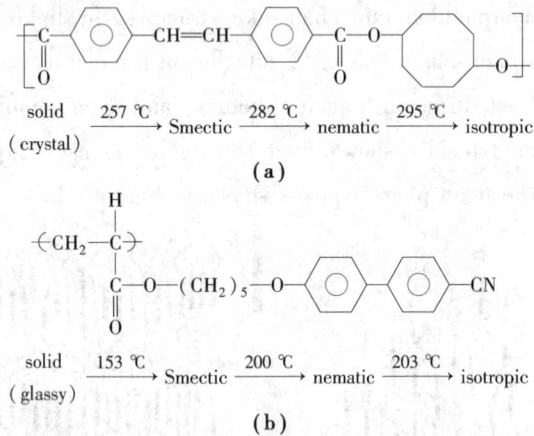

$$\text{solid} \xrightarrow{257\ ^\circ\text{C}} \text{Smectic} \xrightarrow{282\ ^\circ\text{C}} \text{nematic} \xrightarrow{295\ ^\circ\text{C}} \text{isotropic}$$
(crystal)

(a)

$$\text{solid} \xrightarrow{153\ ^\circ\text{C}} \text{Smectic} \xrightarrow{200\ ^\circ\text{C}} \text{nematic} \xrightarrow{203\ ^\circ\text{C}} \text{isotropic}$$
(glassy)

(b)

Figure 8.3　Examples of thermotropic liquid crystal polymers showing multiphase transitions.(a) main-chain polymer; (b) side-chain polymer

From the standpoint of polymer applications, two properties of LCPs are of major interest—the effect of order on polymer melt viscosity, and the ability of the polymer to retain its ordered configuration when cooled down to the solid state.

Among the first polymers observed to exhibit the aforesaid properties of LCPs were copolyesters (I) prepared from terephthalic acid, ethylene glycol, and p-hydroxybenzoic acid.

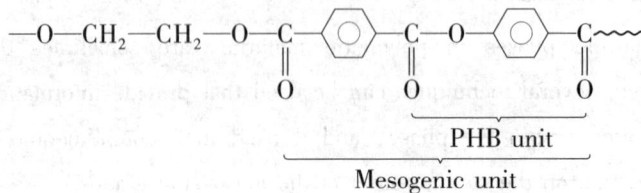

(I)

As the amount of p-hydroxybenzoate (PHB) units is increased, the polymer melt viscosity initially increases, which is expected because of the decreased flexibility caused by incorporation of

220

the "rigid" PHB unit. At levels of 30 mol% PHB, however, the melt viscosity begins to decrease, reaching a minimum at about 60 mol% ~ 70 mol%. This is shown in Figure 8.4(a) at three different shearing rates. Significantly, as the melt viscosity begins to decrease, the melt's appearance also changes from clear to opaque.

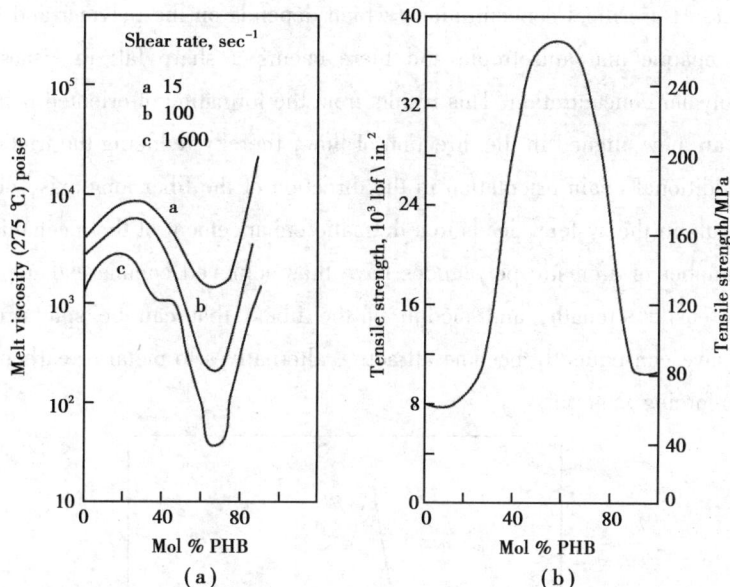

Figure 8.4   Effect of p-hydroxybenzoic acid concentration on (a) melt viscosity and
(b) tensile strength of a terephthalic acid/p-hydroxybenzoic acid/ethylene glycol copolyester

Both the decrease in viscosity and appearance of opaqueness arise from the onset of liquid crystalline morphology, which in turn is due to increased backbone rigidity. The rigid polymeric mesophases become aligned in the direction of flow, thus minimizing the frictional drag. Liquid crystal melts or solutions have, in consequence, lower viscosities than melts or solutions of random-coil polymers. The significance of this effect from the standpoint of polymer processing is obvious: the lower the viscosity, the more readily the polymer can be fabricated into a useful plastic or fiber.

An equally important observation for the above copolyester LCPs is that the ordered arrangement of polymeric mesophases in the melt is retained upon cooling, which is manifested in greatly improved mechanical properties (see Figure 8.4(b)). The liquid crystalline behavior is therefore advantageous from the standpoint of both processing and properties. Thermotropic liquid crystal copolyesters of structures similar to (I) are now available commercially.

Ordered behavior is also observed in solutions of some liquid crystal polymers (lyotropic LCPs). Unlike flexible polymers that assume a random coil conformation in solution, the rigid polymers being rod-like tend to cluster together in bundles of quasi-parallel rods as their concentration in solution in increased. These form domains that are anisotropic and within which there is nematic order of the chains. There is, however, no directional correlation between the domains themselves unless the solutions are shared. When shearing takes place the domains tend to

221

become aligned parallels to the direction of flow, thereby reducing viscosity of the system.

Lyotropic LCPs exhibit quite characteristic viscosity behavior in solution as the polymer concentration in solution is changed. Typically the viscosity follows the trend shown in Figure 8.5. As more and more polymer is added to the solvent, the viscosity increases while the solution remains isotropic and clear. At a critical concentration (which depends on the polymer and the solvent) the solution becomes opaque and anisotropic and there occurs a sharp fall in viscosity with further increase in the polymer concentration. This results from the formation of oriented nematic domains in which the chains are now aligned in the direction of flow, thereby reducing the frictional drag on the molecules. The additional chain orientation in the direction of the fiber long axis, obtained from the nematic self-ordering in the system, leads to a dramatic enhancement of the mechanical properties of the polymer. A number of aromatic polyamides have thus achieved commercial importance because of the very high tensile strengths and moduli of the fibers that can be spun from the nematic solutions. These have consequently become attractive alternatives to metal or carbon fiber for use in composites as reinforcing material.

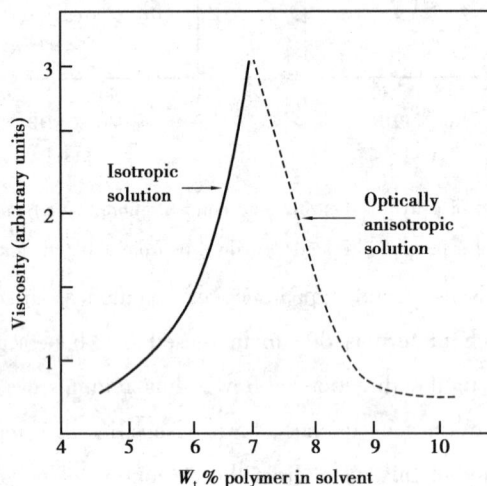

Figure 8.5   Variation of viscosity of solutions of partially chlorinated poly (1,4-phenylene-2,6-naphthalimide) dissolved in solvent mixture of hexamethylene phosphoramide and N-methylpyrrolidone containing 2.9% LiCI, as a function of solution concentration showing transition from isotropic to anisotropic behavior

The most significant of these aramid fibers are:

①Poly(m-phenylene isophthalamide), trade name Nomex

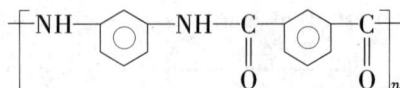

②Poly(p-benzamide) or Fiber B

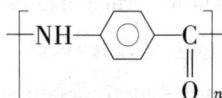

③Poly(p-phenylene terephthalamide), trade name Kevlar

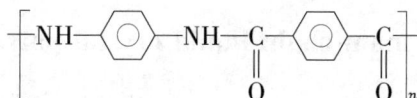

$$\left[NH-\!\!\bigcirc\!\!-NH-\overset{\text{C}}{\underset{\parallel O}{}}-\!\!\bigcirc\!\!-\overset{\text{C}}{\underset{\parallel O}{}}\right]_n$$

The last-named polymer exhibits liquid crystalline phase in sulfuric acid solution. The solution is extruded to form a fiber, resulting in further alignment of the molecules. The product, once the sulfuric acid is removed, is a fiber with a more uniform alignment than could be obtained simply by drawing and thus has much better mechanical properties. Tensile strength of Kevlar, for example, is considerable higher than that of steel, whereas its density is much lower. Although most Kevlar produced is used in tie cord, the polymer also finds use in specialty clothing. Light-weight bulletproof vests are made containing up to 18 layers of woven Kevlar cloth.

LCPs have significantly increased crystalline melting temperatures as a result of the extended chain morphology. In fact, major drawbacks to the type of rigid polymers that exhibit liquid crystalline behavior are that they have a very high melting point—e.g., poly(p-hydroxybenzoic acid) melts at about 500 ℃—and are difficult, it not impossible, to dissolve in the usual organic solvents. This makes them difficult to process, so alternative structures with much lower melting points are more useful.

The melting points of LCPs can be reduced in a number of different ways (schematically represented in Figure 8.6), namely,

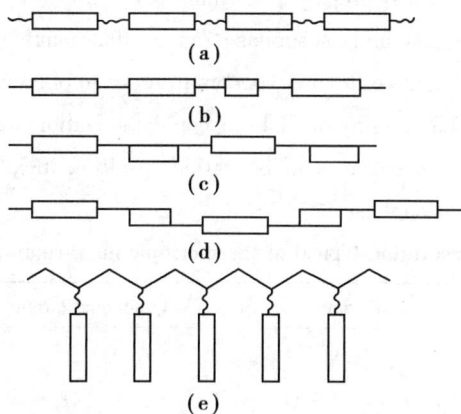

(a)

(b)

(c)

(d)

(e)

Figure 8.6   Schematic representation of several
arrangements of mesogens (□) and flexible spacers (~)
in main chain and side chain of liquid crystalline polymers

①Incorporation of flexible spacer units, in chain to separate the rigid backbone groups (mesogens), which are responsible for the mesophases.

②Copolymerization of several mesogenic monomers of different sizes to give a random and more irregular structure.

③Introduction of kinks in the main chain, such as by using meta substituted monomers or a crankshaft monomer (e.g., 6-hydroxy-2-naphthoic acid).

④Attachment of mesogens to the polymer backbone via flexible spacers.

8.2.1 Thermotropic main-chain liquid crystal polymers

The use of flexible spacers is a popular approach for producing thermotropic main-chain polymer liquid crystals. The mesogenic moiety consists of two cyclic units, normally joined by a short rigid bridging group. These are then linked through functional groups to flexible spacers of varying length that separate the mesogens along the chain and thus reduce the overall rigidity.

The bridging groups are usually multiple bond units, because they must be rigid to maintain the overall stiffness of the mesogens. Ester groups also serve this purpose, particularly when the cyclic units are aryl rings where the conjugation leads to a stiffening of the overall structure (Ⅱ) :

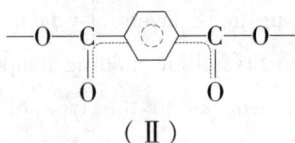

$$ -O-\overset{\overset{\parallel}{O}}{C} -\underset{}{\bigcirc}- \overset{\overset{\parallel}{O}}{C} -O- $$

(Ⅱ)

Many of the examples of thermotropic main-chain LCPs are polyesters that are synthesized by condensation reactions including interfacial polymerizations, or by high-temperature solution polymerizations using diols and diacid chlorides. However, the preferred method is often an ester interchange reaction in the melt. Among the commonly used monomer units are p-hydroxybenzoic acid, terephthalic acid, 2,6-naphthalene dicarboxylic acid, 2-hydroxy-6-naphthoic acid, and 4, 4'-biphenol. The main-chain LCPs prepared in this way, however, tend to be very insoluble polymers with high melting points and mesophase ranges that make them difficult to process. A common approach is thus to introduce flexible spacers in order to obtain more tractable LCPs (Table 8.1). The spacer units are usually introduced by a copolymerization reaction and the proportion of the spacer units relative to the mesogens can be varied resulting in alteration of the melting point and the temperature range of mesophase.

Table 8.1 Chemical constitution typical of thermotropic main-chain-polymer liquid crystals

Cyclic Unit	Bridging Group	Functional Group	Spacer
(benzene ring, $x=1\text{--}3$)	$\overset{\overset{O}{\parallel}}{-C-O-}$	$\overset{\overset{O}{\parallel}}{-C-O-}$	$(CH_2)_n$
(naphthalene, 1,4 1,5 2,6)	$\underset{R}{-C}=N-N=\underset{R}{C-}$	$\overset{\overset{}{-O-C}}{\underset{O}{\parallel}}$	$(CH_2-CHO)_n$ with R
(cyclooctane ring)	—CH=N— ; —N=N— ; $\underset{O}{-N=N-}$	$(CH_2)_n$	$\overset{R}{\underset{R}{-Si-O-}}$
(cyclohexane, H, H)	$\underset{R}{-CH=C-}$	—O—	—S—R—S—

224

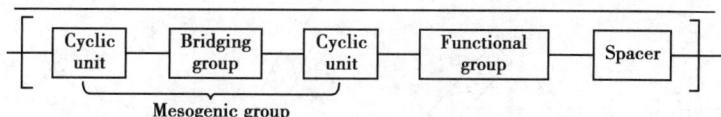

Mesogenic group

The majority of the main chain LCPs having group arrangements as shown in Table 8.1 exhibit a nematic phase after melting, but in some cases small variations in structure can lead to formation of a smectic mesophase. Thus for polyesters with the following structures:

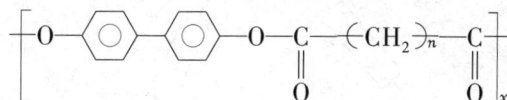

A nematic phase is observed when the number of methylene units (n) in the spacer is odd, but a smectic mesophase results when n is even. Similarly, for polyesters with multiple rings, the phases can be nematic or smectic depending on the orientation of the ester units, e.g., (III) and (IV):

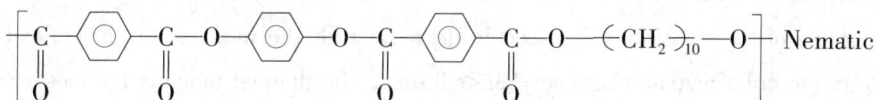

(III)

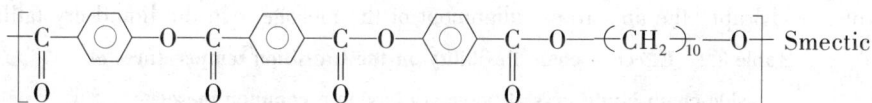

(IV)

The length of the spacer unit has significant effect on the melting point (T_m) and isotropic transition temperature (T_i), and hence on the temperature range ($T_i - T_m$) in which the mesophase is stable.

Polymers with spacers having an even number of CH_2 units usually have higher melting (T_m) and clearing (T_i) temperatures than those with an odd number. This suggests that the spacer length influences the ordering in the liquid crystal phase. As shown schematically in Figure 8.7, it is easier with even-numbered methylene unit spacers to maintain the orientation of the mesogen parallel to the director axis in the all trans zig-zag conformation.

8.2.2   Side-chain liquid crystal polymers

It has been demonstrated that polymers with mesogens attached as side chains can exhibit liquid crystalline properties. The extent to which mesophases can develop in these system is influenced by the flexibility of the backbone chain and whether the mesogen is attached directly to the chain or is separated from the chain by a flexible spacer unit.

The degree of flexibility of the polymer chain to which the mesogens are bonded can affect both the glass transition temperature (T_g) and the mesophase to isotropic phase transition temperature

225

n=7   n=8

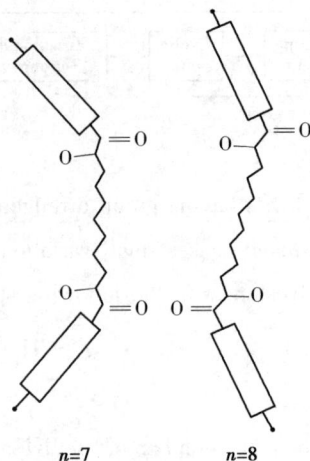

Figure 8.7   Schematic diagram showing the effect of odd and even number of —CH$_2$—

spacer units on the relative orientation of the mesogenic units in a main-chain liquid crystal polymer

(T_i). This is illustrated in Table 8.2 for a number of side-chain LCPs having the same mesogen. The temperatures of the transitions are seen to decrease with the increase in chain flexibility, the latter being in the order methacrylate<acrylate<siloxane. The thermal range of the mesophase (ΔT) is thus the greatest when the chain is most flexible and its conformational changes largely do not interfere with, or disrupt, the anisotropic alignment of the mesogens in the liquid crystalline phase.

Table 8.2   Effect of chain flexibility on the transition temperatures of side-chain liquid crystal polymers having a common mesogen

Polymer	Transitions(℃)	ΔT(℃)
$\begin{array}{c} \text{CH}_3 \\ \mid \\ \text{[CH}_2\text{—C]}_n \\ \mid \\ \text{COOR} \end{array}$	Glassy $\xrightarrow{187}$ Nematic $\xrightarrow{201}$ Isotropic	14
$\begin{array}{c} \text{[CH}_2\text{—CH]}_n \\ \mid \\ \text{COOR} \end{array}$	Glassy $\xrightarrow{160}$ Nematic $\xrightarrow{177}$ Isotropic	17
$\begin{array}{c} \text{CH}_3 \\ \mid \\ \text{[O—Si]}_n \\ \mid \\ \text{CH}_2\text{R} \end{array}$	Glassy $\xrightarrow{142}$ Nematic $\xrightarrow{168}$ Isotropic	26

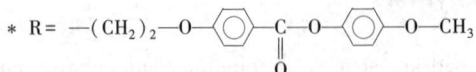

* R= —(CH$_2$)$_2$—O—⟨◯⟩—C—O—⟨◯⟩—O—CH$_3$
$$\qquad\qquad\qquad\qquad\quad \|$$
$$\qquad\qquad\qquad\qquad\quad O$$

The influence of the polymer backbone on the alignment of the sidechain mesogens can be minimized by decoupling the motions of the main chain from those of the mesogens. This can be achieved by introducing long flexible spacer units between the backbone and the mesogen. Structures of this type can be synthesized in a number of ways. One such scheme is shown in Figure 8.8.

226

$$HO\!-\!\bigcirc\!-\!CO_2H \xrightarrow[\text{Cl(CH}_2)_6\text{OH}]{\substack{\text{KOH,KI,H}_2\text{O,}\\ \text{EtOH}}} HO(CH_2)_6O\!-\!\bigcirc\!-\!CO_2H$$

$$CH_2\!=\!CH\underset{CO_2(CH_2)_6O-\bigcirc-CO_2H}{} \xleftarrow[\substack{CO_2H\\ HO-\bigcirc-OH}]{CH_2\!=\!CH\ ,PTSA}$$

(1) SOCl$_2$, DMF

t-Bu $\overset{OH}{\underset{CH_3}{\bigcirc}}$ t-Bu

(2)

$$HO\!-\!\bigcirc\!-\!CN$$

NEt$_3$

$$CH_2\!=\!CH\underset{CO_2(CH_2)_6O-\bigcirc-CO_2-\bigcirc-CN}{}\rightarrow Polymer$$

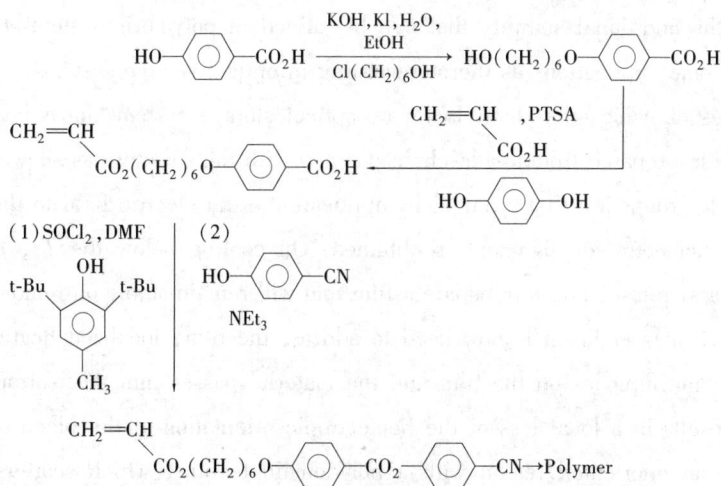

Figure 8.8　A representative scheme of synthesis of side-chain liquid crystal

It is generally observed that as longer spacer units are introduced, the T_g of the polymer is lowered by internal plasticization and the tendency for the more ordered smectic phase to develop is increased. Both these phenomena reflect the known tendency for long side chain to order and eventually to crystallize, when sufficiently long, and this condition persists also in the liquid crystal state.

The ordered state of the mesophase in the aforesaid crystalline polymers is readily frozen and locked into a glassy state if the temperature is rapidly brought down below the T_g and remain stable until heated above T_g again. The phenomenon offers several interesting application possibilities in optoelectronics and information storage. These applications often depend on the ability of the mesogenic groups to align under the influence of an external magnetic or electric field, as discussed below.

As the dielectric constant and diamagnetic susceptibility of many mesogens are anisotropic, the orientation of sidechain LCPs in the nematic state can also be changed by the application of a magnetic of electric field. The parameter of interest is the magnitude of the critical field, which is required to affect transition (Fredericks transition) from the homogeneous to the homeotropic aligned state (see Figure 8.9).

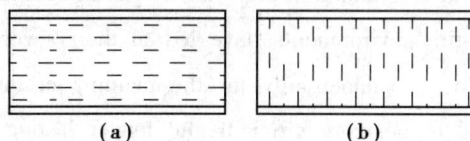

(a)　　　　　(b)

Figure 8.9　Schematic representation of (a) homogeneous and
(b) homeotropic alignment of mesogens in a measuring cell

Whereas the relaxation time for this transition in low-molecular-weight mesogens is of the order of seconds, this may be several orders of magnitude larger in polymer systems due to viscosity effects. Though this makes the use of polymeric liquid crystals less attractive in rapid-response

227

display devices, the additional stability that can be gained in polymeric systems can be distinctly advantageous for some applications as thermorecording in optical storage systems.

The principles of using side-chain LCPs as optical storage systems have been demonstrated using a polymer film prepared from a side-chain polymer with the structure as shown in Figure 8.3b. The mesogenic side groups are first oriented by application of an electric field to the polymer above its T_g, such that homeotropic alignment is obtained. On cooling below the T_g, the alignment is locked into the glassy phase, and a transparent film that will remain stable on removal of the electric field is produced. If a laser beam is now used to address the film, localized heating occurs at the point where the beam impinges on the film and the material passes into the isotropic, disordered, melt state. This results in a local loss of the homeotropic orientation at the place of laser exposure and, on cooling, an unoriented region with a polydomain texture, which scatters light and thus produces a nontransparent spot, forms in the film (see Figure 8.10). Information can thus be "written" onto the film, and can subsequently be erased by simply raising the temperature of the whole film to regain the isotropic, disordered, melt state.

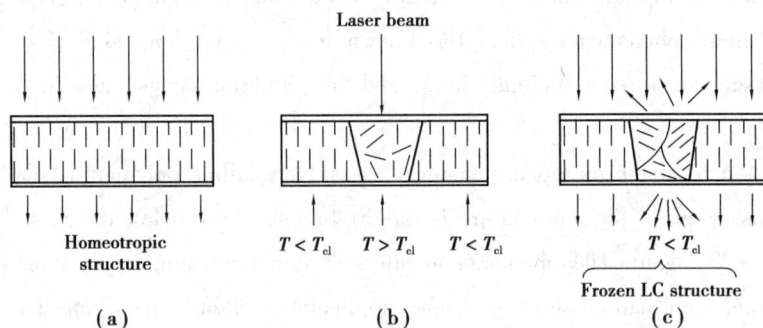

Figure 8.10 Thermal recording using a homeotropically aligned side-chain liquid crystal polymer as a transparent film. A laser beam is used to address the film by producing local heating and disorder, which is subsequently frozen in by cooling

8.2.3 Applications

One of the first applications of LCPs was a range of dual-ovenable cookware of Xydar made by Tupperware, a subsidiary of Dartco Manufacturing. The resistance of LCPs to chemicals and solvents and good performance in hostile environments have led to their several specialized applications. Vectra materials have been used commercially for the molding of formic acid separation tower packings as they have proved to be more efficient and longer lasting than conventional ceramic packing materials. LCPs are also used in such demanding areas as surgical instruments, aircraft and automotive engine systems, fiber optic devices, chemical equipment, and photocopier components.

LCPs are finding use as replacement for epoxy and phenolic resins in electrical and electronic components, printed circuit boards (PCBs), and fiber optics. In these applications, the high mechanical properties, low coefficient of thermal expansion, inherent inflammability, good barrier properties, and ease of processing of LCPs (Vectra in particular) are important.

In fiber optic applications, LCPs can be extruded into a variety of shapes and sizes using conventional extrusion equipment found in typical fiber optic wire and cable production plants.

Having one of the highest flow rates of any polymer and virtually no deformation or shrinkage on molding, LCPs are typically used for precision parts with thin walls and complex shapes. In the electronics industry, the benefits mentioned above—plus the resin's resistance to soldering temperatures of 200~250 ℃—give it an edge over other high-performance plastics in many surface-mounted devices. For example, finely dimensioned electronic components such as SIMM socket with 0.050 in. spacing between pins are typical uses of LCPs.

In several electronic applications, such as connectors and capacitors, LCPs are also being used in preference to high-performance materials, such as PPS, because of their ease of processing, greater impact resistance, and overall lower part cost.

8.3   Conductive Polymers

As computers and sophisticated electronics devices began to move out of their shielded rooms into offices, stores, and homes, it became highly desirable to take advantage of the light weight, low cost, and aesthetics that could be gained by substituting plastics for metal housings for the instruments. Conductive plastics have therefore been increasingly used to provide flexible, lightweight, and moldable parts having good static bleed-off and electromagnetic interference (EMI) shielding properties. A variety of uses of such materials are encountered, ranging from compliant gasketing to rigid housing for business machines[3-6].

Conductive polymers fall into two distinct categories: filled polymers, which are used for a wide range of anti-static and static-discharging applications, and intrinsically conductive polymers, which contain no metals but conduct electricity when chemically modified with dopants.

A number of other polymeric solids have also been the subject of much interest because of their special properties, such as polymers with high photoconductive efficiencies, polymers having nonlinear optical properties, and polymers with piezoelectric, pyroelectric and ferroelectric properties. Many of these polymeric materials offer significant potential advantages over the traditional materials used for the same application, and in some cases applications not possible by other means have been achieved.

8.3.1   Filled Polymers

Polymers can be made to conduct electricity relatively easily by compounding them with high loadings of conductive materials as fillers. Apart from the inherent properties of the fillers, parameters such as concentration, particle form (sphere, flake, fiber), size, distribution, and orientation are deciding factors that influence the properties of filled polymers.

When choosing a filler, the following requirements merit consideration: ①the filler has high conductivity to avoid excess weight; ②it does not impair the physical/mechanical properties of the plastic; ③it is easily dispersed in the plastic; ④it does not cause wear on forming tools (injection molding, extrusion); ⑤it has favorable cost picture; and ⑥it produces good surface structure of the finished product. The commonly used electrically conducting fillers are carbon, aluminum, and steel. The most common metallic conductor, copper, is not used because it oxidizes within the plastic and impairs its physical properties.

Typical dimensions for common fillers are:

①Aluminum flakes: length 1~1.4 mm, thickness 25~40 mm

②Carbon black: spherical, diameter 25~50 mm

③Graphite fiber: length mm~cm, thickness 8 mm

④Steel fiber: length 3~5 mm, diameter 2~22 mm

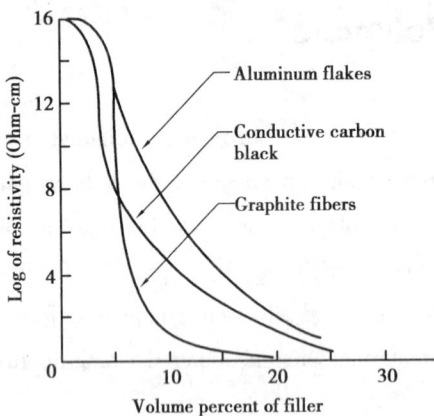

Figure 8.11   Composite resistivity as a function of filler volume loading and filler type

All of them are to be found in a number of qualities with varying prices and conduction properties. Depending on particle form and orientation, there is a certain critical volume concentration at which the resistance decreases, i.e., the conductivity increases, drastically. This is shown in Figure 8.11. At the critical concentration, the filler can form a continuous phase through the matrix in the form of microscopic conductive channels. As shown in Figure 8.11, the specific resistance of filled polymers also depends on the inherent conductivity and particle form of the filler besides its concentration. The critical concentration can be reduced to low levels by using conductive particles that are fibrous in shape. The reduction in critical volume loading is proportional to the magnitude of the fiber's aspect ratio (length/diameter).

It has been shown that even extremely small concentrations of additives can make plastics conductive if they are in the form of conductive fibers with length to diameter (L/D) ratio of 100 or more. The striking difference between the use of chunky fragments and fibrous materials in their effect on conductivity can be seen from the diagram in Figure 8.12.

If one loads a plastic with 5 percent by volume of 6 ml size in a random distribution (Figure 8.12(a)), there will be, on the average, a 6-mil gap to the nearest sphere. Heat or electrons flowing through such a matrix would thus cross alternate paths of about equal lengths in the two media. The picture, however becomes quite different if the same 5 percent of material is dispersed as 1 ml diameter and 100 ml long fibers. It is inevitable that at random orientations they will touch

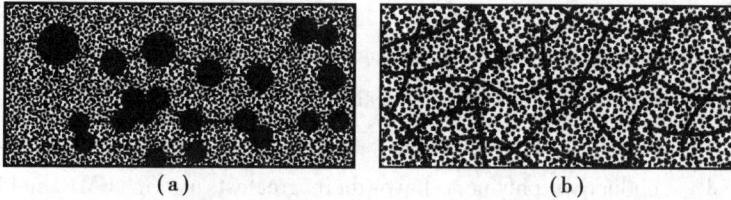

Figure 8.12   Comparison of typical flow path through composites using
the same volume percentage of material as spheres and fibers (L/D=100).
(a) Chunky particles;long path through plastic; (b) fibers;short gaps in plastic

one or more of their nearest neighbors, as shown in Figure 8.12(b), thereby providing an almost continuous path through the composite along the conductive fibers. Fibers will therefore be more effective in lowering the electrical resistivity of plastic and increasing the thermal conductivity than chunky fragments.

The above concept led to the creation of metalloplastics—a family of conductive plastics in which the conductivity is great enough to make the material resemble metal both electrically and thermally. To serve as a practical engineering material, metalloplastics should have electrical resistivities of less than 1 Ω and thermal conductivities of at least a factor of ten higher than those of normal plastics. The importance of high thermal conductivity in plastics is being recognized as the automobile business tries hard to use more and more plastic parts to reduce the weight of the automobile. With conventional plastics having poor thermal conductivity, the problem of getting heat into the part to form it and then getting the heat out again results in cycle times of the order of minutes, not seconds. This implies an increase of one to two orders of magnitude in the number of tools necessary to make the same number of parts. So, for the auto industry, an order of magnitude increase in thermal conductivity could be a very significant advance.

Metalloplastics with electrical resistivities as low as 0.01 Ω and up to 100-fold higher thermal conductivity than ordinary plastics have been developed by the addition of a few percent of metal and/or metallized glass fibers (L/D of the order of 100/1) to plastics. Use of such materials can have very significant effects on molding cycle times, uniform heating and cooling rates, and heat transfer rates in the final product. Whereas, thermal conductivity is proportional to the concentration of conductive fillers and is increased even by such low concentrations of the fibers that one fiber does not touch its neighbors, the electrical resistivity is not significantly modified until an almost continuous path is available through the conductive fibers. Plastics can thus be developed that are improved in thermal conductivity but can be used for electrical insulation or resistive heating. Suitably filled polymers are thus used to drain off heat in pressure switches as well as in polymeric tapes intended for self-regulating, resistive heating of water pipes, railroad switches, etc.

Filled conductive polymers are 10~12 orders of magnitude more conductive than unfilled polymers but are still several orders of magnitude lower than copper. Carbon-black filled polymers are the most common. Fillers other than carbon blank include finely divided metal flakes and fibers, metallized glass

231

fibers, and metallized inorganics such as mica.

Filled conductive polymers used for packaging include polycarbonate, polyolefin, and styrenics incorporating fillers such as carbon, aluminum, and steel flakes and fibers. A polycarbonate/ABS blend introduced by Bayer is 4% aluminum filled and suited to many screening functions.

The electrically conductive polymers have their greatest use in EMI-shielded casings and protective housings of electronic and telecommunications equipment, where they have rapidly substituted metal because of their low weight and easy workability. Steel fiber is mostly used in filled conductive polymers intended for EMI-shielding applications, which will be discussed below.

Conductive rubber has been used in a number of applications. In most cases, silicone rubber is used because of its greater resistance to temperature, UV light, oxygen, corrosive gases, chemicals, and solvents as compared to organic rubber. Conductive rubber is used in tires to leak off static electricity. Ethylene-propylene conductive rubber is used as a covering in cables to reduce high voltage problems. In all these cases, carbon black is generally used as the filler, one reason being that it also acts as a reinforcing filler, increasing the strength and tear resistance of the product. In the electronics industry, conductive rubber finds use as a connector for liquid crystal displays in electronic digital watches.

8.3.2 Inherently Conductive Polymers

Interest in electrically conducting polymers began in the mid-1960s with the suggestion of Little in his theoretical studies that some organic materials could become superconductors. The first example that has been extensively studied is tetrathiafulvalene-tetracyanoquinodimethane and its derivatives. The first covalent polymer exhibiting the electronic properties of a metal was polymeric sulfur nitride, $(SN)_x$, which attracted much attention in the mid-to late 1970s.

In 1977, the first covalent organic polymer, polyacetylene $(CH)_x$, that could be doped through the semiconducting to the metallic range, was reported. Another significant breakthrough occurred in 1980 with the discovery that PPP could be doped to conductivity levels quite comparable to those in polyacetylene. This polymer is the first example of non-acetylene hydrocarbon polymer that can be doped to give polymers with metallic properties.

Polyacetylene has been investigated much more extensively than any other conducting polymer and has served as a prototype for the synthesis and study of a large number of other conjugated, dopable organic polymers. Of these, the greatest research efforts are devoted to polypyrrole, PPS, PPP, polythiophene, and polyaniline.

In a nondoped state, the basic polymers have low conductivity. The two polyacetylene conformations cis and trans are in the semiconductor range; PPP is a good insulator.

The synthesis of the polyacetylene powder has been known since the late 1950s, when Natta used transition metal derivatives that have since become known as Ziegler-Natta catalysts. The characterization of this powder was difficult until Shirakawa and coworkers succeeded in synthesizing lustrous, silvery, polycrystalline films of polyacetylene (which has become known as "Shirakawa" polyacetylene) and in developing techniques for controlling the content of cis and trans isomers:

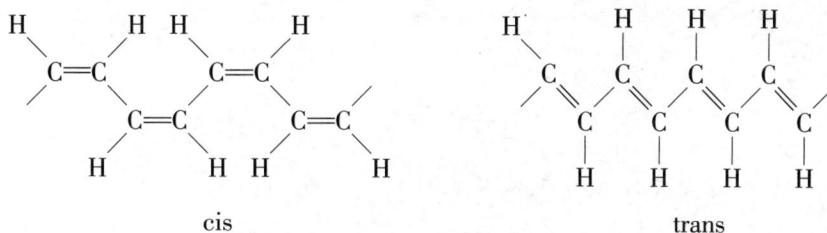

cis   trans

The PPP structure has all the characteristics required of a potential polymeric conductor, but it has proved difficult to synthesize a high-molecular-weight material. One method is the polycondensation, but this only yields oligomeric material, and even this is insoluble.

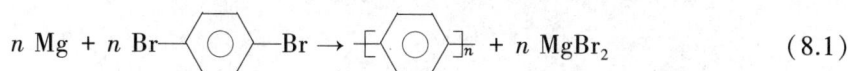

$$n \text{ Mg} + n \text{ Br} \!-\!\!\left\langle\bigcirc\right\rangle\!\!-\! \text{Br} \rightarrow \left[\left\langle\bigcirc\right\rangle\right]_n + n \text{ MgBr}_2 \tag{8.1}$$

The polymer is very stable and can withstand temperatures up to 450 ℃ in air without degrading. It is an insulator in the pure state but can be both n- and p-doped using methods similar to those for polyacetylene. However, as PPP has a higher ionization potential it is more stable to oxidation and requires strong p-dopants. It responds well to AsF_5, with which it can achieve conductivity levels of 10^2 Ω.

8.3.3 Conduction mechanism

Again, poly (acetylene) is taken as an example to illustrate principles of conduction mechanisms in conducting polymers.

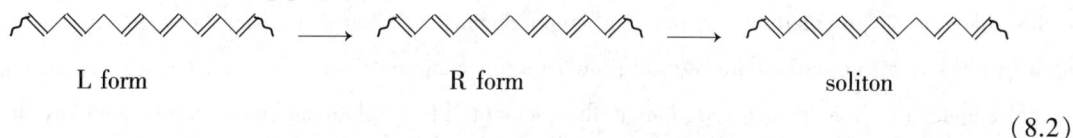

L form   R form   soliton

$$(8.2)$$

As a necessary consequence of the asymmetry of the poly (acetylene) ground state, two equivalent polyene chains R and L are interconverted through the intervention of a mobile charge carrier, a soliton. The soliton is a mobile charged or neutral defect, or a "kink" in the poly (acetylene) chain that propagates down the chain and thus reduces the barrier for interconversion.

The charge carrier in n-doped (negatively charged) poly(acetylene) is a resonance-stabilised polyenyl anion of approximately 29~31 CH units in length, with highest amplitude at the centre of the defect. This description is backed up by chemical shifts found in the ^{13}C NMR spectrum, which indicate increased charge density at the centre of such anions. Also advanced theoretical calculations (MNDO) are in agreement with this description.

So how does a soliton manage to travel from one end of a sample to the other? This can be explained by the bipolaron hopping mechanism(Figure 8.13).

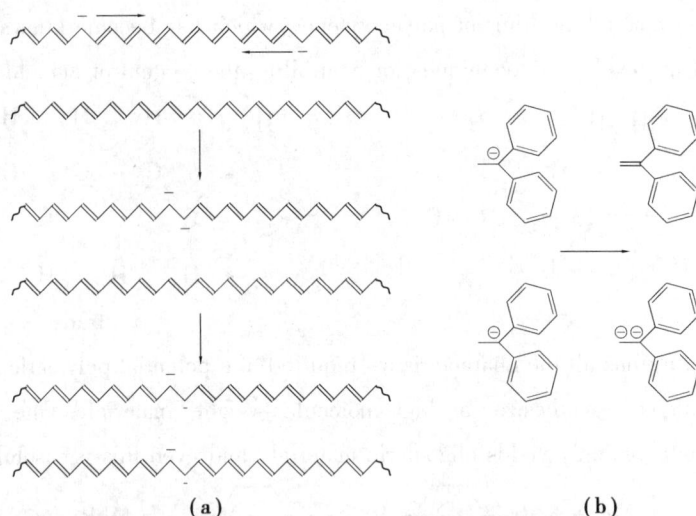

(a)   (b)

Figure 8.13   Illustration for the bipolaron hopping mechanism

Most experimental data is consistent with a rate-limiting interchain charge transport. If short chains are involved, conduction in poly(acetylene) requires some mechanism for transfer of charge from one chain to another (e.g., intersoliton hopping). Current theories for interchain charge transfer centre around the Kivelson "percolation" mechanism. For n-type solitons (Carbanions) this involves electron transfer from the anion to a neutral soliton (radical) in a neighbouring chain in an isoenergetic process.

However the lattice distortion associated with interpolaron hopping (hopping between a radical carbanion and a neutral polyene chain) is thought to present inaccessible energy barriers. A solution to this dilemma is provided by organic carbanion chemistry. Charge transfer between carbanions often involves a disproportionation equilibrium between a neutral/dianion pair (ion triplet) and a radical anion pair. It is recognised that at the moment this mechanism most closely matches the charge transport energetics.

8.3.4　Applications

Conducting polymers (CPs) are an exciting new class of electronic materials, which have attracted an increasing interest since their discovery in 1977. They have many advantages, as compared to the non-conducting polymers, which is primarily due to their electronic and optic properties. Also, they have been used in artificial muscles, fabrication of electronic device, solar energy conversion, rechargeable batteries, and sensors. This study comprises two main parts of investigation. The first focuses conducting polymers (polythiophene, polyparaphenylene vinylene, polycarbazole, polyaniline, and polypyrrole). The second regards their applications, such as supercapacitors, light emitting diodes (LEDs), solar cells, field effect transistor (FET), and biosensors.

234

(1) Supercapacitors

Supercapacitors have been focused on the development of new modified electrode materials with improved performance. The electrode materials for supercapacitors have been classified into three categories: transition metal oxides, high-surface carbons, and conducting polymers[7].

The development of hybrid electric vehicles and the fast-growing market of the portable electronic devices have prompted an increasing and urgent demand for environment friendly high-power energy resources. Supercapacitors are also called electrochemical capacitors or ultracapacitors. Because of their pulse power supply, long cycle life, simple principle, and high dynamic of charge propagation, they have attracted much interest. In order to form fast charging energy-storage devices of intermediate specific energy, they are designed to fill the gap between batteries and capacitors. They have a high market potential regarding both the hybrid electric vehicles and the pure electric vehicles as to improve the regenerative braking and provide larger acceleration. The capacitances are delivered in mF and μF quantities. Capacitors have been developed to give hundreds to thousands of Farads. These are usually known as supercapacitors, or ultracapacitors, and are initially constructed from carbons of high surface area.

(2) Ligth emitting diodes (LED)

Burroughes et al. studied the polymer light-emitting diodes (PLEDs), which have been become the topic of intense academic and industrial research. PLEDs based on PPV are now coming out as commercial products. When compared to inorganic or organic materials for LEDs, the main advantages of the polymer electroluminescence (EL) devices are their fast response times, process ability, the possibility of uniformly covering large areas, low operating voltages, and the many methods were applied to fine-tune their optical and electrical properties by varying the structure. At present, only green and orange LEDs meet the requirements of commercial use, even though all three primary colors (red, green and blue) have been exhibited in LEDs. Polymers in the electronics industry overtake their long established passive roles as insulating and encapsulating materials to more active new applications. They can be also designed for microlitographic applications.

(3) Solar cell

Fossil energy usage causes serious environmental problems. It is necessary to look for clean and renewable energy resource, such as solar energy, which is called the really green energy, having nearly unlimited supply capability and being widely distributed all over the earth. In spite of the fact that the direct photovoltaic energy conversion in matters of magnitude is more energy efficient than any of those indirect sources, the global use of photovoltaic (PV) is only emerging at a slow pace. The issue behind is that the cost of PV modules based on traditional PV technology is still too high to be afforded by common energy consumption.

The use of polymeric materials in the design and fabrication of low cost organic electronic devices, photovoltaic devices, or plastic electronics, has received much attention. When comparing the organic technology to the silicon-based photovoltaics (PV) , the two very different technologies are complementary in many ways. Organic photovoltaics (OPVs) offer low cost solution processing, flexible substrates, low thermal budget and a very high speed of processing. When organic solar cells are compared to the established inorganic solar cell techniques, a significant disadvantage of the organic solar cells is the low overall power conversion efficiency. There are several factors influencing the efficiency of OPVs, such as the structure of the polymer, the morphology of the film, the interfaces between the layers (organic/metal, organic/organic), and the choice of electron acceptor. To realize exciton dissociation in organic PV cell, the electron donor/acceptor approach is an effective way. The photoactive layer in a polymer solar cell should at least consist of two constituents, which are electron donor (donor or D) and electron acceptor (acceptor or A), respectively. The widely used donor constituents are mainly the conjugated polymers, such as polyphenylene vinylene (PPV) , polythiophene (PT) , polyfluorene (PF) or their derivatives.

However, other donor materials with much lower bandgap are present; for example, the copolymers consisting of the segments of thiophene, fluorene, pyridine, and so on. The electron acceptor materials heavily used are usually C60 or its derivatives, the inorganic nanoparticle acceptor, such as ZnO and CdSe, and the conjugated polymers with cyano group having strong electron affinity.

Solution-processed bulk-heterojunction (BHJ) solar cells were first reported by Friend et al. with all polymer active thin films and followed by Heeger et al. in 1995 with a polymer/fullerene composite. The general device architecture is fabricated in sandwich geometry (Figure 8.14). Transparent, conducting electrodes such as ITO are used as the substrate which can be structured by chemical etching. Poly (ethylene-dioxythiophene) doped with polystyrene sulfonic acid (PEDOT∶PSS) is coated as the interfacial layer to facilitate the hole injection/extraction. Vacuum evaporation and solution processing techniques are two of the most commonly used thin film preparation methods to coat the blended active layers[8].

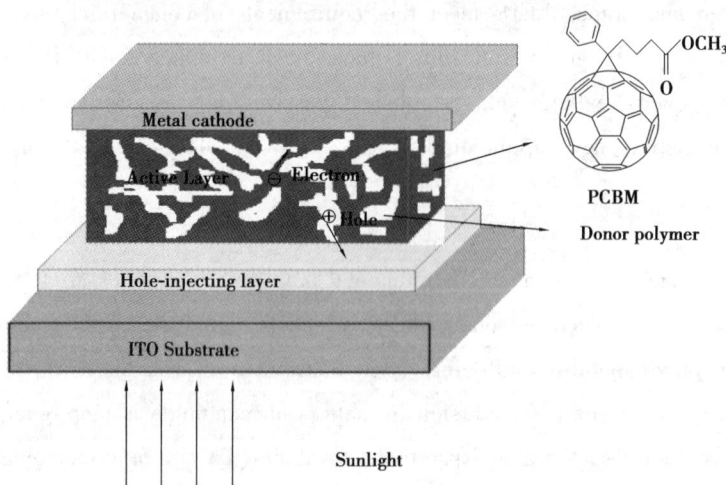

Figure 8.14   Graphic representation of the device architecture of BHJ solar cells

The general mechanistic picture of an organic solar cell comprises the following four steps∶ ①light absorption and exciton formation, ②exciton migration, ③charge separation and ④charge transport and charge collection. Because of the high absorption coefficient and low carrier mobility, thin electroactive films with a thickness of around 100 nm are used. Upon photo-absorption, an exciton is formed due to the low dielectric constants of organic semiconductors. The exciton has to migrate to the interface between donor and acceptor domains, where charge separation occurs. Since the lifetime of an exciton is short, the migration distance of the exciton is therefore limited (~10 nm). Thus, controlling the morphology in BHJ solar cells is crucial. Further device studies have suggested that the utilization of an electron or hole blocking layer is beneficial to achieving high efficiency in the final solar cell. Due to the versatility in fine tuning the absorption windows of the resulting polymers, tandem solar cells become an attractive option to harvest more solar energy in the whole solar spectral range.

A bandgap of 1.1 eV (1 100 nm) is capable of absorbing 77% of the solar irradiation, but semiconducting polymers have bandgaps higher than 2 eV (620 nm) and can only harvest about 30% of the solar photons. In order to harvest as much as solar spectrum, conjugated polyaromatic polymers, especially those with thiophene moieties, have been developed to provide polymers with tunable bandgaps, from 1.0 to 2.0 eV. In the meantime, the tuning of energy levels, as well as the solubility, is crucial to achieve high PCE values in OPV devices via the optimization of solar cell parameters. Two strategies are used for these polymers: to tune both solubility and energy levels, various substituents are attached to the polymer backbones so that morphology and performance of the electroactive layer can be optimized. Several structures of promising building blocks for potential high performance materials are presented in Figure 8.15. Stille coupling polycondensation is widely used for thiophene-containing polymers, which can tolerate numerous functional groups. The Suzuki coupling reaction is applied for preparing polymers with phenyl repeating units.

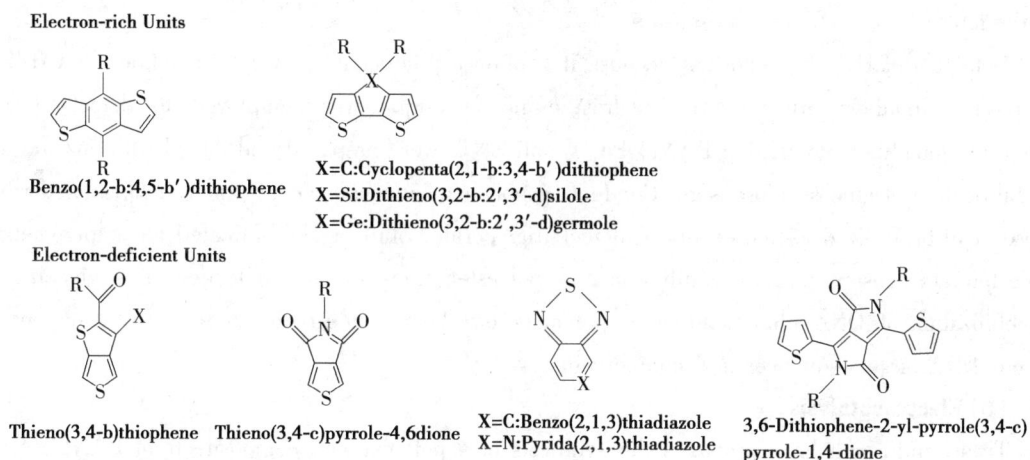

Electron-rich Units

Benzo(1,2-b:4,5-b′)dithiophene

X=C:Cyclopenta(2,1-b:3,4-b′)dithiophene
X=Si:Dithieno(3,2-b:2′,3′-d)silole
X=Ge:Dithieno(3,2-b:2′,3′-d)germole

Electron-deficient Units

Thieno(3,4-b)thiophene   Thieno(3,4-c)pyrrole-4,6dione

X=C:Benzo(2,1,3)thiadiazole
X=N:Pyrida(2,1,3)thiadiazole

3,6-Dithiophene-2-yl-pyrrole(3,4-c)
pyrrole-1,4-dione

Figure 8.15   Several promising building blocks toward high performance materials

(4) Field effect transistors (FET)

Conducting polymers'advantages over conventional materials, such as silicon and germanium, include low cost and ease of processing. Organic or polymer-based semiconductors have been applied to fabricate field-effect transistors (FETs) since 1983. There have been many ongoing efforts to form organic or polymer-based FETs. Organic or polymer based transistors have already found their application, such as in smart pixels and sensors.

Various approaches involve several techniques, including solution processed deposition, spin coating and printing, electro-polymerization, vacuum evaporation, etc. Other techniques (soft lithography, self assembly, and Langmuir-Blodgett) have been also applied to the fabrication of polymer based FETs and have been used to deposit conducting polymer or organic semiconductors.

Liu et al.[9] produced FET device poly(3,4-ethylenedioxythiophene) working as the source/drain/gate electrode material and polypyrrole acting as the semiconducting layer. Poly (vinyl pyrrolidone) K60 (PVP-K60), an insulating polymer, operates as the dielectric layer. The construction of the device follows a number of steps. First, a layer of aluminum 2 000 Å thick is deposited on the silicon dioxide wafer and patterned with UV lithography to form the contact pads. Next, PEDOT/PSS is printed on the Al gate pad at a substrate to form the gate electrode. Then, the PVP-K60 dispersion in water is dispensed onto the gate. The third printing step was to dispense PPy onto the PVP-K60 to form the active layer. Finally, the top source and drain electrodes made of PEDOT/PSS are printed onto the top of the PPy active layer and also extended to the Al source/drain contact pads under the same conditions which are used for printing the gate electrode.

(5) Biosensors[10]

Recently, conducting polymers have attracted much interest in the development of biosensors because they act as excellent materials for immobilization of biomolecules and rapid electron transfer for the fabrication of efficient biosensors.

Detection of H_2O_2 is important because it is often a product in enzymatic reactions. PANI/PS composite nanofibers prepared by electrospinning technique were employed to detect H_2O_2. Composite nanofibers containing PANI, Fe_3O_4 and CNTs were prepared and doped with enzyme for the fabrication of glucose biosensors. Conducting polymer nanocomposites, when encapsulated with lipase, can be utilized as biosensors to detect triglyceride. Shin et al. fabricated an amperometric cholesterol biosensor using polyaniline-coated polyester films for the detection of triglycerides. Immobilization of DNA onto conducting polymers has been extensively studied for detection of various DNA target sequences and microorganisms.

(6) Electrocatalysis

Tiwari and Singh have proposed the synthesis of a polymer nanocomposite from PANI/PAA/MWCNTs by an in situ chemical polymerization method. The nanocomposite thus formed has improved catalytic, electrochemical and electrical behavior. Huang et al. fabricated PANI/Au composite nanotubes as electrodes for the oxidation of NADH. Zhao and his coworkers fabricated poly (N-isopropyl acrylamide)-grafted multiwalled carbon nanotubes onto a PniPAm-modified substrate for bioelectrocatalysis of NADH. PPy/Cobalt porphyrin and PANI/Cobalt porphyrin composite nanorods displayed good electrocatalytic properties of oxygen reduction in neutral electrolytes.

As a result, conducting polymers have been considered for important materials in microelectronics applications, electrocatalysis, fuel cell electrodes, light emitting diodes, biosensor microelectrodes, reinforced composites, biomedical applications and etc.

8.4 Ion exchange resins

Ion exchange resins are polymers that are capable of exchanging particular ions within the polymer with ions in a solution that is passed through them. This ability is also seen in various natural systems such as soils and living cells. The synthetic resins are used primarily for purifying water, but also for various other applications including separating out some elements.

In water purification the aim is usually either to soften the water or to remove the mineral content altogether. The water is softened by using a resin containing Na^+ cations but which binds Ca^{2+} and Mg^{2+} more strongly than Na^+. As the water passes through the resin the resin takes up Ca^{2+} and Mg^{2+} and releases Na^+ making for a "softer" water. If the water needs to have the mineral content entirely removed it is passed through a resin containing H^+ (which replaces all the cations) and then through a second resin containing OH^- (which replaces all the anions). The H^+ and OH^- then react together to give more water[11].

8.4.1 Fundamentals of ion exchange

Synthetic ion exchange materials based on coal and phenolic resins were first introduced for industrial use during the 1930s. A cation exchange resin with a negatively charged matrix and exchangeable positive ions (cations) is shown in Figure 8.16. Ion exchange materials are sold as spheres or sometimes granules with a specific size and uniformity to meet the needs of a particular application. The majority are prepared in spherical (bead) form, either as conventional resin with a polydispersed particle size distribution from about 0.3 mm to 1.2 mm (50-16 mesh) or as uniform particle sized (UPS) resin with all beads in a narrow particle size range. A few years later resins consisting of polystyrene with sulphonate groups to form cation exchangers or amine groups to form anion exchangers were developed (Figure 8.17). These two kinds of resin are still the most commonly used resins today.

A variety of functional groups have been added to the condensation or addition polymers used as the backbone structures[12]. Porosity and particle size have been controlled by conditions of polymerization and uniform particle size manufacturing technology. Physical and chemical stability have been modified and improved. As a result of these advances, the inorganic exchangers (mineral, greensand and zeolites) have been almost completely displaced by the resinous types except for some analytical and specialized applications. Synthetic zeolites are still used as molecular sieves.

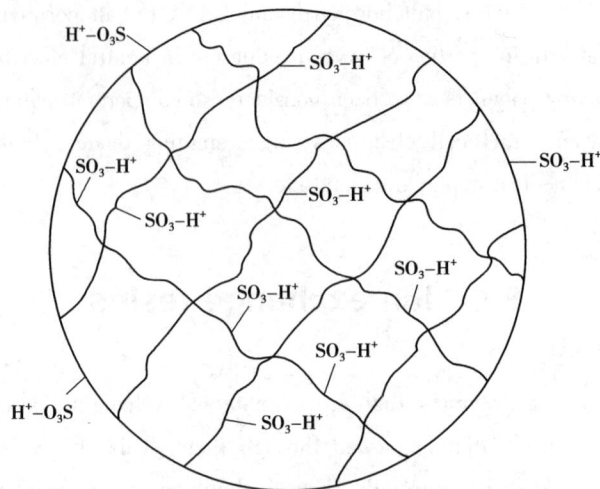

Figure 8.16   Cation Exchange Resin Schematic Showing
Negatively Charged Matrix and Exchangeable Positive Ions

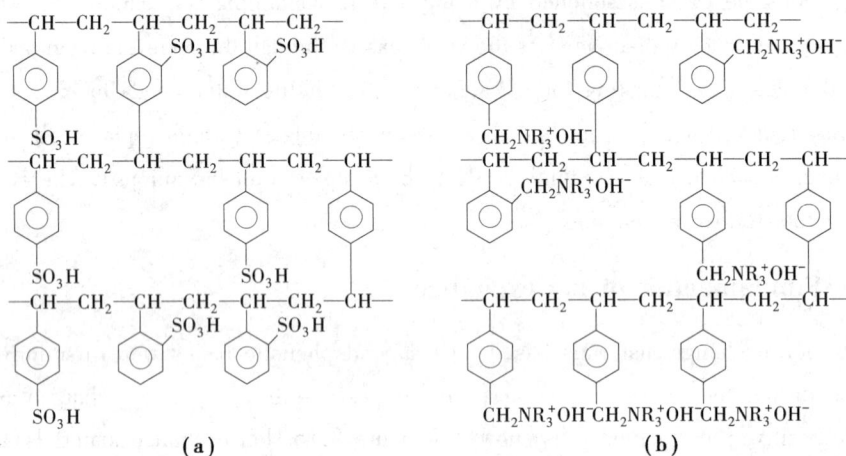

(a)   (b)

Figure 8.17   Some examples of ion exchange resins (a) A strongly acidic sulphonated polystyrene
cation exchange resin; (b) A strongly basic quaternary ammonion anion exchange resin

A corresponding list for amine based anion exchangers is:

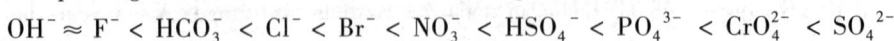

$$OH^- \approx F^- < HCO_3^- < Cl^- < Br^- < NO_3^- < HSO_4^- < PO_4^{3-} < CrO_4^{2-} < SO_4^{2-}$$

Suppose a resin has greater affinity for ion B than for ion A. If the resin contains ion A and ion B is dissolved in the water passing through it, then the following exchange takes place, the reaction proceeding to the right (R represents the resin):

$$AR + B^{n\pm} \Longleftrightarrow BR + A^{n\pm} \tag{8.3}$$

When the resin exchange capacity nears exhaustion, it will mostly be in the BR form.

The manufacture of ion exchange resins involves the preparation of a cross-linked bead copolymer followed by sulfonation in the case of strong acid cation resins, or chloromethylation and the amination of the copolymer for anion resins as shown in Figure 8.18.

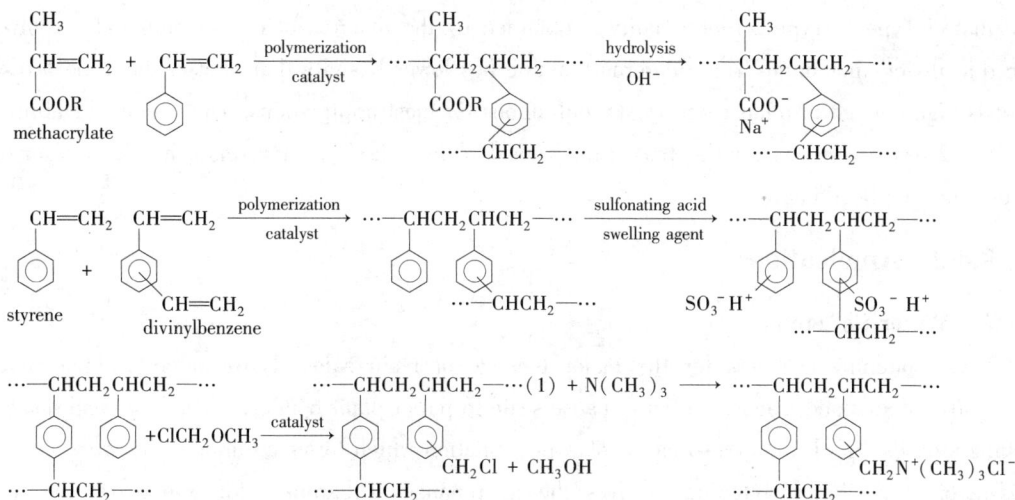

Figure 8.18　Ion exchange resin structure and synthesis

①Cation Exchange Resins: Weak acid cation exchange resins are based primarily an acrylic or methacrylic acid that has been cross-linked with a di-functional monomer (usually divinylbenzene (DVB)). The manufacturing process may start with the ester of the acid in suspension polymerization followed by hydrolysis of the resulting product to produce the functional acid group.

Weak acid resins have a high affinity for the hydrogen ion and are therefore easily regenerated with strong acids. The acid-regenerated resin exhibits a high capacity for the alkaline earth metals associated with alkalinity and a more limited capacity for the alkali metals with alkalinity. No significant salt splitting occurs with neutral salts. However, when the resin is not protonated (e.g., if it has been neutralized with sodium hydroxide), softening can be performed, even in the presence of a high salt background.

Strong acid resins are sulfonated copolymers of styrene and DVB. These materials are characterized by their ability to exchange cations or split neutral salts and are useful across the entire pH range.

②Anion Exchange Resins: Weak base resins do not contain exchangeable ionic sites and function as acid adsorbers. These resins are capable of absorbing strong acids with a high capacity and are readily regenerated with caustic. They are therefore particularly effective when used in combination with a strong base anion by providing an overall high operating capacity and regeneration efficiency.

Strong base anion resins are classed as Type 1 and Type 2. Type 1 is a quaternized amine product made by the reaction of trimethylamine with the copolymer after chloromethylation.

The Type 1 functional group is the most strongly basic functional group available and has the greatest affinity for the weak acids such as silicic acid and carbonic acid, that are commonly present during a water demineralization process. However, the efficiency of regeneration of the resin to the hydroxide form is somewhat lower, particularly when the resin is exhausted with monovalent anions, such as chloride and nitrate. The regeneration efficiency of a Type 2 resin is considerably greater than that of Type 1. Type 2 functionality is obtained by the reaction of the styrene-DVB copolymer with dimethylethanolamine. This quaternary amine has lower basicity than that of the Type 1 resin, yet it is high enough to remove the weak acid anions for most applications. The chemical stability of the Type 2 resins is not as good as that of the Type 1 resins, the Type 1 resins being favored for high temperature applications.

8.4.2　Applications

(1) Water Softening

Water softening accounts for the major tonnage of resin sales. Hard waters, which contain principally calcium and magnesium ions, cause scale in power plant boilers, water pipes and domestic cooking utensils. Hard waters also cause soap precipitation which forms an undesirable gray curd and a waste of soap. Water softening involves the interchange of hardness for sodium on the resin. Typically, hard water is passed through a bed of a sodium cation exchange resin and is softened.

(2) Dealkalization

Many industrial processes require that hardness and alkalinity be removed from a raw water before the water is used in the process. Two main processes involving ion exchange are used for dealkalizing:

①Dissolved solids are removed to the extent of the alkalinity in the raw water by passing the raw water through a bed of weak acid cation resin in the hydrogen form. The 100 percent utilization of regenerant acid that is characteristic of this process decreases operating costs and greatly minimizes the waste disposal problem. A weak acid cation resin creates no free mineral acidity in the effluent when regenerated at a level of not more than 105~110 percent of the theoretically required amounts for the cations picked up.

②Chloride anion dealkalizing involves passing the raw water through a Type 2 anion exchange resin that is in the chloride form to remove alkalinity.

(3) Demineralization

Ion exchange demineralization is a two step process involving treatment with both cation and anion exchange resins. Water is passed first through a column of strong acid cation exchange resin that is in the hydrogen form (RH^+) to exchange the cation in solution for hydrogen ions.

(4) Condensate Polishing

Single or mixed bed ion exchange resins are used in deep bed filter demineralizers for reduction of particulate matter and dissolved contaminants in utility power plant condensates.

(5) Ultra Pure Water

Ultra pure water (UPW) is essential to the proper fabrication of integrated circuit boards in the semiconductor industry. As the degree of integration becomes increasingly more complex, the semiconductor industry requires higher levels of water purity. Single beds, mixed beds and also reverse osmosis are used in the production of ultra pure water.

(6) Nitrate Removal

Ion exchange is used for the removal of nitrates from nitrate polluted waters. Strong base anion exchange resins operating in the chloride ion form (salt solution regenerated) have been successfully used for this service.

(7) Waste Treatment

Radioactive. Radiation waste systems in nuclear power plants include ion exchange systems for the removal of trace quantities of radioactive nuclides from water that will be released to the environment. The primary resin system used is the mixed bed.

(8) Chemical Processing-Catalysis

Since ion exchange resins are solid, insoluble (but reactive) acids, bases, or salts, they may replace alkalis, acids and metal ion catalysts in hydrolysis, inversion, esterification, hydration or dehydration, polymerization, hydroxylation and epoxidation reactions. The advantages of ion exchange resins as catalysts include easy separation from the products of reaction, repeated reuse, reduction of side reactions and lack of need for special alloys or lining of equipment.

Furthermore, there are many other applications for ion exchange resins.

Collection of Exercises

1. What are functional polymers?
2. 简述液晶的类型,以及高分子液晶的应用领域。
3. 溶致型液晶高分子和热致型液晶高分子有哪些主要区别?
4. 结构型导电高分子和复合型导电高分子的差别是什么?
5. 简述离子交换树脂的用途。
6. 简述离子导电聚合物的构成及导电原理。
7. 如何使离子型高分子树脂再生?
8. 导电高分子除具有金属和半导体的特性外,还具有高分子结构的哪些特性?
9. 什么是掺杂? 为什么掺杂后的共轭高聚物的电导率可大幅度提高?
10. What are the basic characteristics of the structure of inherently conducting polymers?
11. What is the advantage of a conductive polymer over a copper wire?

REFERENCES

[1] K.Horie, M.Baron, R.B. Fox, et al., Definitions of terms relating to reactions of polymers and to functional polymeric materials, Pure and Applied Chemistry, 76(4): 889-906, IUPAC, 2004.

[2] Manas Chanda, Salil K. Roy, Industrial Polymers, Specialty Polymers, and Their Applications, CRC Press, 2008.

[3] Roncali J, Conjugated Poly (Thiophenes)-Synthesis, Functionalization, and Applications, Chemical Reviews, 92(4): 711-738, 1992.

［4］ Feast Wj, Tsibouklis J, Pouwer Kl, Groenendaal L, Meijer Ew, Synthesis, Processing and Material Properties of Conjugated Polymers, Polymer, 37(22): 5017-5047, 1996.

［5］ Hide F, Diazgarcia Ma, Schwartz Bj, Heeger Aj, New Developments in the Photonic Applications of Conjugated Polymers, Accounts of Chemical Research, 30(10): 430-436, 1997.

［6］ Leclerc M, Faid K, Electrical and Optical Properties of Processable Polythiophene Derivatives: Structure-Property Relationships, Advanced Materials, 9(14): 1087-1094, 1997.

［7］ Murat Ates, Tolga Karazehir and A. Sezai Sarac, Conducting Polymers and their Applications, Current Physical Chemistry, 2(3): 224-240, 2012.

［8］ Tao Xu, Luping Yu, How to design low bandgap polymers for highly efficient organic solar cells, Materials Today, 17(1): 11-15, 2014.

［9］ Liu, Y.; Varahramyan, K.; Cui, T. Low-voltage all-polymer field effect transistor fabricated using an inkjet printing technique, Macromolecular Rapid Communication., 2005, 26, 1955-1959.

［10］ Tapan K. Das, Smita Prusty, Review on Conducting Polymers and Their Applications, Polymer-Plastics Technology and Engineering, 51: 1487-1500, 2012.

［11］ http://www.nzic.org.nz/ChemProcesses/water/13D.pdf, Ion Exchange Resins.

［12］ Wheaton RM, Lefevre LJ, DOWEX Ion Exchange Resins-Fundamentals of Ion Exchange, 陶氏化学公司技术资料.